DIY Solar Energy

A Step-by-Step Guide with Tips and Tricks for Installing a Home Photovoltaic System

© Copyright 2023 by John Patterson

© Copyright 2023 by John Patterson- **All rights reserved**.

This Book is provided with the sole purpose of providing relevant information on a specific topic for which every reasonable effort has been made to ensure that it is both accurate and reasonable. Nevertheless, by purchasing this Book, you consent to the fact that the author, as well as the publisher, are in no way experts on the topics contained herein, regardless of any claims as such that may be made within. As such, any suggestions or recommendations that are made within are done so purely for entertainment value. It is recommended that you always consult a professional prior to undertaking any of the advice or techniques discussed within.

This is a legally binding declaration that is considered both valid and fair by both the Committee of Publishers Association and the American Bar Association and should be considered as legally binding within the United States.

The reproduction, transmission, and duplication of any of the content found herein, including any specific or extended information, will be done as an illegal act regardless of the end form the information ultimately takes. This includes copied versions of the work, both physical, digital, and audio, unless express consent of the Publisher is provided beforehand. Any additional rights reserved.

Furthermore, the information that can be found within the pages described forthwith shall be considered both accurate and truthful when it comes to the recounting of facts. As such, any use, correct or incorrect, of the provided information will render the Publisher free of responsibility as to the actions taken outside of their direct purview. Regardless, there are zero scenarios where the original author or the Publisher can be deemed liable in any fashion for any damages or hardships that may result from any of the information discussed herein.

Additionally, the information in the following pages is intended only for informational purposes and should thus be thought of as universal. As befitting its nature, it is presented without assurance regarding its prolonged validity or interim quality. Trademarks that are mentioned are done without written consent and can in no way be considered an endorsement from the trademark holder.

Table of Contents

Benefits of Photovoltaic (PV) Solar Systems 5
The Three Basic Types Of Technology For Pv 5

Chapter 1 - Basic Types of Solar Panels: Monocrystalline, Polycrystalline, and Thin-Film 17

Chapter 2 - Types of PV System Connections 39
- Grid-Tied Solar Systems 40
- Equipment Components Of A Grid-Tied SolarSystem 41
- Grid-Tied Inverters (Gti) 41
- Power Meters (Net Metering Type) 42
- Off-Grid Solar Systems 42
- Battery Banks 44
- DC Disconnect Switch 46
- Off-Grid Inverters 46
- Hybrid Solar Systems 47

Chapter 3 - Major Components of a Solar PV Rooftop System 49
- Solar PV Modules and Mounting Support Hardware 49
- Dc And Ac Disconnect Switches 51
- Solar Pv System Grid-Tied Inverters 53
- Off-Grid Inverters 54
- Dual Micro-Inverters 61
- Other Components Needed for Your PV Solar Rooftop System 66

Chapter 4 - How to Determine the Optimal Size of Your System 69
- Step 1: Determine Your Power Consumption Requirements 69
- Step 2: Load-Sizing Procedure 73
- Step 3: Size Your Battery Storage System 73
- Step 4: Determine the Average Sun Hours Available Per Day 81
- Step 5: Size the Optimal PV System for Your Home 90

Chapter 5 - How to Calculate the Annual Output of a System and the Optimal Number of Panels 93
- Module Power Rating 94
- Module Temperature Factor 94
- Particulate Build-Up Factor 94
- System Wiring And Module Output DifferenceFactor 95
- Inverter Conversion Losses Factor 95
- Module Tilt Angle Factor 95
- Module Compass Direction/Azimuth Factor 96
- Solar Irradiation Index Factor 97

Final Calculations of Solar Energy Output 98

Links to Online Energy Output and System Size Calculators for PV Grid-Connected Systems 101

Chapter 6 - Government Incentives, Rebates, Subsidies, Grants, Tax Credits, and Private Leasing Programs 103

Chapter 7 - Environmental Benefits of PV Solar Systems 109

Chapter 8 - Maximize the Benefits from Your PV Solar Energy Rooftop Installation 111

Refrigerators 111

Air Conditioners 111

Electric Water Heaters, Electric Clothes Dryers, and Portable Electric Space Heaters 112

Home Lighting Fixtures 113

Automatic Light Controllers and Dimmers 119

Double-Pane Windows 120

Insulation 120

Weather Stripping 121

Location and Use of Windows for Optimal Natural Ventilation 121

Chapter 9 - Where to Find Professionally Certified Solar Installers in Your Area .. 123

Chapter 10 - Photographic Step-by-Step Guide on How to Install a Complete PV Solar Rooftop System 127

Site Evaluations 127

Prepare the Roof for Mounting the Solar Panels 132

Mounting the Solar Panels 143

Installing the Electrical Components and Controls 151

Conclusions and Suggestions 165

Glossary of Terms 167

Benefits of Photovoltaic (PV) Solar Systems

Most municipalities permit homeowners or do-it-yourself enthusiasts to install a PV rooftop solar system; often a simple inspection and approval by a qualified electrician or certified solar installer will be all you need.

They're fast to install. Usually an installation can be completed in a week—even by a DIYer—after all the required equipment and materials are on site. Energy independence is an advantage that has strong appeal for many homeowners, and PV solar systems can reduce vulnerability to power loss from grid blackouts or power outages. You should see immediate savings on your electric bills and begin recovering your investment right away.

PV solar systems are also flexible; you can start small and add more PV panels later on. With solar rooftop systems, everything is modular.

Leasing is available in most areas. If you have limited cash flow, this option can provide an easy way to get a PV solar system installed on your rooftop. It's also easy to obtain financing, if required; most US banks have ample experience financing solar PV rooftop systems. In addition, government cash rebates, tax exemptions, and other financial incentives are often available, and these can substantially reduce the cash investment you need to make. PV systems have a relatively short payback period that can be as short as six or seven years—or even less, with rebates and other government incentives.

Your solar panel system should have a long and useful life. A warranty of 25 years from module manufacturers is standard. Systems can be installed on both flat and sloped roofs, and even on vertical surfaces. They provide myriad long-term environmental benefits and require almost zero maintenance.

Lastly, they immediately increase the value of your home, often by more than the cost of the PV solar system itself.

This is a formidable list of advantages to installing a PV solar system in your home. The only "negative" factor is the substantial investment or capital cost of the PV system, but this is generally more than offset by the advantages, financial and otherwise. The guidelines in this book will make it easy for you to calculate the financial and economic aspects of whichever PV solar system you choose for your home. I'll also help you perform a simple cost-benefit analysis of your project or proposed PV system. Reading this book will put you in control.

The Three Basic Types Of Technology For Pv

SOLAR PANELS OR MODULES:

- Monocrystalline silicon panels
- Polycrystalline silicon panels
- Thin-film technology (in panels or other formats)

The best-known and most common of these is the polycrystalline silicon panel, and the third type, "thin film," has up to four different versions, meaning that there are actually up to six kinds of PV technology to choose from. This might seem confusing at first, but this book will make it easy for you to determine which PV technology type is best for your particular needs.

Here are a few more photographs that illustrate the versatility of solar panel applications and installations.

This book will help you understand the various PV solar technologies and enable you to choose the most advantageous system for your home and budget. It will also give you valuable tips on how to choose a contractor, if this is the route you decide to take. It will show you how to calculate the electric power you can produce using any given system and how much money your chosen system can save on your monthly electric bills.

In Chapter 1, I will outline the main features, advantages, and disadvantages of each available technology type so that you can choose from the diverse range of options and components.

In Chapter 2, I will discuss and illustrate three different options for connecting or not connecting to the electric utility grid: "grid-tied," "off-grid," and "hybrid."

In Chapter 3, I will illustrate and summarize all the major components of a PV solar rooftop system.

In Chapter 4, I will show you how to determine the optimum size of your proposed PV solar rooftop system.

In Chapter 5, I will explain how to calculate the annual energy output of any given PV solar rooftop system, and I'll give you two simple ways to calculate the optimum number of solar panels needed to produce the amount of energy you will require under your particular set of circumstances and in your specific location. Chapter 5 will also show you an easy way to calculate how much money your PV solar system can save you compared to your existing monthly utility bills over a 12-month period.

In this photo, we can see the solar PV system is being installed at the same time as the installation of a new roof.

Here, we can see a do-it-yourself crew of three installing mono panels on the aluminum rails of an array that aligns very close to the edge of the roof just above the eaves trough.

A close-up shot of the corner of a mono panel at the end of an array.

Here we see the installer tightening a mid-bracket between two poly panels and securing them to the railings of the support framework.

Numerous solar arrays with over 50 panels mounted on a flat condo rooftop with a tilt of about 20 degrees indicating that this installation is in a location with a latitude of about 20 degrees.

The installer is securing this thin film array onto a very steep roof flush with the roof surface.

A low-level rancher with an array of 11 panels mounted flush to the roof surface.

In Chapters 4 and 5, you will also learn to calculate how long it will take to recover your initial investment—normally only six or seven years, depending on the amount of solar irradiance available throughout the year at your location, and depending on government subsidies, rebates, and other incentives you may qualify for, as explained in Chapter 6.

In Chapter 7, I will summarize the environmental benefits of residential and commercial PV solar systems and show you how to calculate the annual amount of CO_2 emissions that can be prevented from entering the atmosphere with any solar PV system.

In Chapter 8, you will learn many useful and effective measures and complementary mini-projects that will maximize the benefits to be gained with a solar PV project for your home. These measures should not only save you money in the long run, they will also reduce the required size and cost of the solar PV system needed to power 100 percent of your home's requirements. Lots of good ideas here!

In Chapter 9, I will show you how to find a qualified and registered solar installer in your area—or even a DIY coach, who can help you to save money by doing the installation yourself.

In Chapter 10, I will include a detailed, step-by-step photographic procedure on how to install a complete PV solar rooftop system.

In Chapter 11, I will summarize the main conclusions from previous chapters and, using this book as a guide, provide you with some helpful basic recommendations on how you might best proceed with plans to execute your PV solar project.

Solar Rooftop Systems: An Introduction

This is your handbook to planning your own PV solar rooftop system. I'll help you develop a customized action plan for a successful project that will benefit you and your family for decades to come. Photovoltaics, or PV technology, continue to be used in different ways and in new devices all the time: for powering parking-lot and highway lighting; for schools in remote villages with no access to the utility grid; for residential, commercial, and industrial applications; for electric vehicle charging stations; for remote microwave transmission stations; and even for boats, small power tools, and other devices.

Predictions of a solar revolution become more common with every passing year, and in some countries it's already becoming a reality. In Germany, for example, PV solar–installed capacity in 2012 increased by 7.6 Gigawatts to a total of over 35 Gigawatts by late 2013. This represents approximately 3.5 percent of the nation's electricity production. Some market analysts believe this figure could reach 25 percent by 2050. But the government has even higher goals than those predicted by market analysts. Germany announced an official goal of producing 35 percent of electricity from renewable sources by 2020—and 100 percent by 2050! These are formidable national goals, and some would say they are overly optimistic. But the trend is definitely upward around the world.

This phenomenon has been created in large part due to feed-in tariff (or FiT) schemes. These government financial incentive programs—in Germany and in other countries worldwide—have for decades provided subsidies to homeowners in order to encourage and stimulate the growth of the PV solar energy infrastructure.

Other countries that have been following Germany's lead in developing PV solar energy as an important renewable power source include England, the USA, India, Spain, Italy, Australia, France, Canada, Japan, and numerous countries in Southeast Asia. The expansion of the PV solar energy industry has short-and long-term environmental benefits, but there are more powerful economic and political factors behind the growing universal government support for solar energy. The political element usually involves developing solar power in order to decrease future dependency on carbon-based fuel imports, such as diesel and liquefied natural gas (LNG), and to reduce national expenditures on these imports. The goal is to achieve independence from foreign control over energy supplies.

Two important long-term economic trends have also provided incentives for more solar. These trends include, on the one hand, the year-by-year reduction in manufacturing costs and the market price index for PV solar panels, and on the other hand, the increasing annual cost of electricity produced by the national- grid utility companies that use carbon-based fuels, a finite natural resource.

Crude-oil prices on the world market go up or down in any given year, but the five-year average trend has generally been upward over the long term. Between 2015 and '16, it is well known that there was a huge drop in crude oil prices on the world market and this was reflected, to some degree, by decreases in gasoline and fuel oil prices in most countries. However, this did not affect in any major way the cost of electricity to homeowners because the utility's electricity rates to the public are not so directly tied to global crude oil prices, being that the amortized annual cost of the capital equipment used in oil burning generating stations is proportionately so much greater than the cost of the crude oil consumed in these gigantic plants. In any case, the utilities are reluctant or very slow to pass on any savings to a significant degree. Utility rates may decrease slightly for a while but they can be expected to continue to increase over the long term. The net result is that global crude oil prices are of limited consequence to the solar PV residential market.

These two trends will eventually lead to what is known as "grid parity," the point at which the cost of PV solar power to the consumer or end user is equal to the cost of electricity from the national electric grid, usually measured in terms of a cost figure per kilowatt hour, i.e.: "$/kWh" in North America or "€/kWh" in Europe. Grid power in Europe has traditionally been much more expensive than in North America.

An array of nine large mono panels installed parallel to a rooftop has a very minor slant of about 15 degrees, indicating that this location is likely near 15 degrees latitude, probably in the southern United States.

Two DIY installers start to install a large number of poly panels on a low-tilt roof of white asphalt tiles. You will notice that the aluminum rails of the solar system support structure have already been completedwithout any additional tilt.

Four arrays of five poly modules, each installed with two arrays on the main house and two equal-size arrays on the garage. We can see that the roof has a tilt of about 30 degrees and the arrays are mounted with an additional tilt of about 15 degrees for a total tilt of about 45 degrees, indicating that this house islikely located near the 45th degree of latitude to face the sun at 90 degrees for maximum energy output. This is explained in detail in Chapter 5.

Three arrays of 12 poly modules, each that are ground-mounted on three terraces in steep terrain. Whenyour rooftop space is insufficient for your solar power requirements, this type of ground-mounted supplementary PV installation may be a good solution if the extra space is available somewhere behind your house.

In many countries, grid parity has already become a reality and the trendsleading to grid parity are generally considered to be irreversible. In countries where conventional electric energy from carbon-based power plants is still relatively cheap, the incentive to install solar is usually less than in countries closer to grid parity, but these trends are present worldwide.

These political and economic factors are further reinforced by the environmental benefits of using PV solar. Photovoltaics are a totally clean source of electric energy, a stark contrast to conventional energy sources that burn highly toxic carbon-based fuels (diesel oil, coal, or liquid natural gas) to power large electric generators in big buildings characterized by tall, wide-diameter chimneys. During combustion, these conventional carbon-based fuels emit large amounts of carbon dioxide—or CO_2—that are released into the atmosphere and contribute directly to the greenhouse effect that causes global warming. These CO2 gas emissions are quantified in terms of tons of CO2. In Chapter 7, I'll show you how to calculate the savings in CO_2 emissions you can achieve with any PV solar rooftop system for any given time period.

Accurate statistics are available that show how much conventional carbon-based fuel is consumed in the production of a given amount of electricity, measured in kWh. Statistics are easily calculated as to how many tons of CO_2 are emitted by carbon-based fuels while generating electricity, i.e.: for every MWh (1,000 kWh) of power. Therefore, if we know the amount of power produced by a PV solar system in kilowatt hours or megawatt hours, we can calculate exactly how many tons of CO_2 gas emissions are prevented from entering the atmosphere by virtue of using renewable photovoltaic energy instead of conventional carbon-based fuels.

Of all electricity-producing plants, coal-fired generators produce the highest level of CO_2 emissions. Solar PV farms and rooftop PV solar systems produce zero CO_2. The ability to precisely measure CO_2 emissions enables regulatory bodies such as the United Nations (UN) to establish international treaties and national limits for CO_2 emissions. This has led to an international system of carbon credits, which enables us to determine how much CO_2 (in tons) a given PV system prevents from entering the atmosphere.

Chapter 1 - Basic Types of Solar Panels: Monocrystalline, Polycrystalline, and Thin-Film

There are many factors you'll want to take into account when choosing a solar (PV) system for your home. In this chapter, I will provide the basic information about the different types of solar panels for home use (monocrystalline, polycrystalline, and thin-film).

PV cells are fabricated using semiconductors as light-absorbing layers that convert the energy of photons into electricity without causing any noise or air pollution. When photons enter the cell through a transparent contact, they're absorbed by the semiconductor, thus creating electron-hole pairs. A special layercalled a "junction" in the body of the device provides the electric field that separates and gathers the generated electrically charged carriers, which are then collected by wires attached to the solar cell.

Multiple cells are strung together in a protective package to form a PV module. Then many modules are interconnected to form an array. Balance ofSystem (BOS) components are the final element of a working PV system.

To help you better understand what kind of solar panels make sense for yourproposed system, there's a simple recommendation you can keep in mind that will save time and help you make the best decisions for your home. It can be summarized as follows: If the homeowner wants to obtain the maximum power from his or her available rooftop space, then monocrystalline or polycrystalline solar modules will be the best choice. Which of these alternatives you choose will depend on several factors, as explained later in this chapter. Only under special circumstances and for certain special applications would I recommend that you consider thin-film solar cells.

Once you have a pretty good idea about the space issue and other specificconditions at your location, then the above criterion will determine which sections of the book you'll want to focus on.

Let's start with the most common types of solar panels on the market and listtheir benefits and disadvantages. Then we'll look at a few typical scenarios in which certain types of PV solar modules would be better than others.

Crystalline Silicon (c-Si) (Includes all monocrystalline andpolycrystalline PV cells)

Between 85 and 90 percent of the world's photovoltaics are based on some variation of silicon. In recent years, about 95 percent of all PV solar system shipments by US manufacturers to the residential sector were crystalline siliconsolar panels.

The silicon used in PV solar cells takes many forms. The main difference between monocrystalline silicon cells and polycrystalline silicon cells is in the purity of the silicon, and this depends on the manufacturing process. But what does silicon purity really mean? The more perfectly aligned the silicon molecules are, the more efficient the solar cell will be at converting solar energy(sunlight) into electricity (the photoelectric effect).

The efficiency of solar panels goes hand-in-hand with the level of silicon purity in the cells that make up the panels, or modules. But the processes used to enhance the purity of silicon are complex and expensive. However, the energy conversion efficiency rating of the PV solar panel shouldn't be your primary concern. As you'll soon discover, cost and space efficiency are the key factors for most people in selecting the optimum PV solar system, as well as the number of panels or modules and the power or number of watts per panel.

What's important to remember right now is that the total number of panels used for a PV rooftop installation, multiplied by the power rating in watts peak (Wp) of each panel, divided by 1,000, will give you the total power of the entire system in kWp. The significance of the term "peak" will be explained later.

Crystalline vs. Thin-Film

The thin-film type of PV panel is often the most cost-effective choice for tropical applications or for hot, desert-like climates. Thin-film panels can produce up to 30 percent more energy per year in hot climates as compared to crystalline panels. Many consumers tend to analyze the different types of photovoltaic panels and make their decision based on the estimated kilowatt hours produced annually by the system, not on the watts peak (Wp) value. But you'll see later in this chapter that there are other factors you'll need to take into account besides the potential annual kWh produced.

Let's take a moment to clarify what the Wp value really means. The watts peak figure, given for any photovoltaic panel, is not the output from that panel. The actual output depends on the surface temperature of the panel, the intensity of light reaching the panel (the Insolation Index), the average daily "sun hours" at the site location, and the angle (tilt to the horizontal plane) and east-west orientation (azimuth angle) of the array to the sun, as well as the type and model of panel being considered. Watts peak (Wp) is a figure obtained under controlled laboratory conditions that do not occur in real life. It's the maximum output of the panel, under perfect conditions. It's important to point out that these laboratory "Wp" conditions apply only to crystalline panels, not to thin-film panels.

The website energyinformative.org/solar-panel-comparison is one of the most comprehensive and up-to-date solar panel comparison resources on the Internet. The solar cell comparison table below provides a quick overview that will also be helpful. This chart lists the most commonly used solar cell technologies today, with their basic characteristics and summary specifications.

A compact array of six mono panels installed flush with the roof surface—a very typical mounting.

Thin-film solar cells include four sub-categories, defined as follows:

- Amorphous silicon (A-Si)
- Cadmium telluride (CdTe)
- Copper indium gallium selenide (CIS/CIGS)

Organic (or polymer) solar cells (in early stages of research and development)

Before proceeding with your purchase, be aware that whichever PV solar technology you select, you must be diligent and careful to analyze the reputation of the available module manufacturers as well as their standard warranty terms.

It really pays to read the fine print. You should look beyond efficiency ratings to the financial health and long-term prospects of the companies issuing these warranties. We'll explore this issue in more detail below.

Solar Cell Comparison Chart by Technology Type: Monocrystalline, Polycrystalline, and Thin-Film					
DESCRIPTION	**MONOCRYSTALLINE**	**POLYCRYSTALLINE**	**AMORPHOUS SILICON**	**CDTE**
Typical module efficiency	16–18 percent	14–16 percent	7–8.1 percent	13–14 percent
Best research cell efficiency	25 percent	21 percent	13.4 percent	19 percent

Area in meters squared (M2) required for 1 kWp or equivalent (Space efficiency)	6–9 M2	8–9 M2	13–20 M2	11–13 M2
Typical length of warranty (see warning note below)	25 years	25 years	10–15 years	10–15 years
Lowest price (subject to change)	0.75 $/W	0.62 $/W	0.69 $/W	0.57 $/W
Temperature resistance	Performance drops 10–15 percent at high temperatures.	Slightly less temperature resistant than monocrystalline or thin film.	Tolerates high ambient temperature.	Relativ impact perform
Additional details	Oldest cell technology and widely used.	Less silicon waste in the production process. Well-proven technology. Widely used.	Tend to degrade over time crystalline-based solar pan and stability issues. The ca CdTe is highly toxic.	Low availability in the m many regions. A-Si prove for small electronic devic portable calculators.

Note: Thin-film systems include the three columns on the right side of this table.
The data above is obtained from the website of the National Renewable Energy Laboratory at www.nrel.

A close-up shot of a mono cell, the basic building block of a monocrystalline PV panel.

Monocrystalline Silicon Solar Cells

Invented in 1955, monocrystalline, meaning single crystal, is the original PV technology. Commercially, monocrystalline entered the market in the late 1970sand is well known for its durability. The useful life of monocrystalline solar modules is about 35 years. They're consistent in performance and reliability.

Single-crystal modules are composed of cells also called "wafers" cut from acylindrical block of continuous crystal. Monocrystalline cells may be fully round, or they may be trimmed into other shapes. Because each cell is cut from asingle crystal, it has a uniform color that's dark blue or dark gray with a hint ofblue.

Solar cells made of monocrystalline silicon (mono-Si), also called single- crystalline silicon cells (single-crystal-Si), are quite easily recognizable by their even coloring and uniform look, indicating high-purity silicon.

Monocrystalline and polycrystalline solar panels are easy to differentiate.

Polycrystalline solar cells are perfectly rectangular, with no rounded edges. Theytypically have a speckled bright blue color.

Since they're made out of the highest-grade silicon, monocrystalline solar panels have the highest efficiency rates, running from 16 to 18 percent, typically,over the last decade. In recent years, SunPower has produced the highest- efficiency solar panels on the US market to date. Their E20 series provides panelconversion efficiencies of up to 20.1 percent. SunPower more recently releasedtheir X-series, with an impressive efficiency rating of 21.5 percent.

In addition, monocrystalline silicon solar panels are the most space efficient.

Since these solar panels yield the highest power outputs, they also require the least amount of space compared to any other type of solar panel. They produceup to four times the amount of electricity compared to most thin-film solar panels for the same panel surface area. This is good to remember if your available rooftop space is somewhat insufficient.

Furthermore, monocrystalline solar panels last the longest. Most solar panel manufacturers put a 25-year warranty on their monocrystalline solar panels. Themonocrystalline silicon modules manufactured in the early 1980s are still functioning, according to many accounts.

Lastly, these particular solar panels tend to perform better than similarly rated polycrystalline solar panels under low-light conditions. For many climates, this also is an important consideration.

However, monocrystalline solar panels also have some disadvantages. They're the most expensive of all photovoltaic solar module types. From a financial standpoint, a solar panel made of polycrystalline silicon might be a better choice for some homeowners, depending on the desired power output andavailable rooftop space.

Another downside is that if the solar panel is partially covered with shade,dirt, or snow, the entire array circuit can break down. Consider getting micro- inverters instead of the old standard conventional central inverters if you think shading or other obstructions such as snow might be a problem. Micro-inverters will make sure that the entire solar array isn't affected by shade hitting only one or two of the panels in the array. The critically important topic of whether to choose micro-inverters or central or string inverters is discussed in more detail in Chapter 3.

The Czochralski process used to produce monocrystalline silicon is technically complex and requires very expensive equipment. This process creates large cylindrical ingots. Four sides are cut out of the ingots, and the four corners are rounded to make silicon wafers. Therefore, a significant amount of the original silicon ends up as waste. This is another factor that contributes to thehigher cost of monocrystalline panels.

Finally, be aware that monocrystalline solar panels tend to be more efficient than the alternatives in hotter climates. The performance of all solar panel types,including monocrystalline, will suffer where average temperatures are much higher, but this is truer for polycrystalline solar panels. However, for most American homeowners, temperature won't be a major concern.

Polycrystalline Silicon Solar Cells

Polycrystalline cells are made from a similar silicon material as that of monocrystalline cells, except that instead of being grown into a large single crystal, the raw material of polycrystalline cells is melted and poured into a mold. This mold forms the silicon into a square block that is then cut into squarewafers, which waste less space and material than the round-cornered single- crystal mono-wafers. As the material cools, it crystallizes in an imperfect manner, forming random crystal boundaries and irregular color textures.

The efficiency of energy conversion for polycrystalline is slightly lower thanmonocrystalline cells of equal surface area. This means that the surface area per watt of energy produced by a polycrystalline module is greater than that of a monocrystalline or single-crystal module. The poly cells look different from the mono-or single-crystal cells, as their surface has a jumbled or speckled look withmany variations of bright blue colors. In fact, they're often quite beautiful.

Some companies have developed alternatives to the traditional molds, such as ribbon growth and growth of crystalline film on glass. Most crystalline silicon technologies yield similar results with high durability. Twenty-five-year warranties are common for crystalline silicon modules.

The silicon used to produce crystalline modules is derived from silicon sand. It's the second most common element on earth, so why is it so expensive? Well, in order to produce the photovoltaic effect, the silicon must be purified to an extremely high degree. Such pure semiconductor grade silicon is very expensive to produce. It's also in high demand in the electronics industry because it's the base material for computer chips and many other devices. Crystalline solar cells are about as thick as a fingernail; therefore, an entire solar PV module uses a relatively large amount of silicon.

Regardless of the technique used in growing the crystals, the construction of finished modules from crystalline silicon cells is generally the same. The most common construction method involves laminating the cells between a tempered glass front and a plastic backing, and then applying a clear adhesive similar to that used in automotive safety glass between the layers. The sheets are rectangular and all have approximately the same dimensions, usually about 1 meter wide by 1.8 meters long. The laminated sheets are then framed around the four edges with a sturdy aluminum "U" channel. The finished panel is about five to six centimeters thick and weighs approximately 30 kilos, which can be easily handled by one installer. That being said, the mounting of a module onto its rooftop support framework is usually handled by two installers, as shown in many of the photographs in Chapter 10.

Two poly arrays on two different slopes of the same roof.

A small array of seven poly modules.

The first solar panels based on polycrystalline silicon, also known as polysilicon (p-Si) or multi-crystalline silicon (mc-Si), were introduced to the market in the 1980s. Unlike monocrystalline-based solar panels, polycrystallinesolar panels do not require the Czochralski process.

Ultimately, the process used to make polycrystalline silicon cells and modules is simpler and costs less compared to monocrystalline. In addition, the amount of waste silicon in the manufacturing process of polycrystalline cells ismuch less compared to monocrystalline. This reduces the production costs ofpolycrystalline cells.

Polycrystalline solar panels tend to have slightly lower heat tolerance than monocrystalline. In practical terms, this means that they perform at a slightly lower level compared to monocrystalline solar panels in high ambient temperatures, for example over 40 degrees C (104 degrees F). Very high ambienttemperatures can slightly shorten the panels' useful lifespan. However, this effect is minor, and most homeowners don't need to take it into account unless they live in a very dry, hot desert.

As compared to monocrystalline, the efficiency rating of polycrystalline- based solar panels is typically 14 to 16 percent, which is due to lower siliconpurity. Its space efficiency is lower as well. You generally need to cover a nominally larger surface to generate the same amount of electrical power as youwould generate with a solar panel made of monocrystalline silicon.

Another drawback to polycrystalline panels is their aesthetic appeal.

Monocrystalline and thin-film solar panels can sometimes be more aestheticallypleasing since they have a more uniform look as compared to the irregular, speckled blue colors of polycrystalline silicon, as shown in the photograph below.

However, this doesn't mean that every monocrystalline solar panel performs better than polycrystalline silicon panels. You must take into account your unique situation and its needs when comparing the two.

Thin-Film Technologies

Imagine if a PV cell was made with a microscopically thin deposit of silicon instead of a thick wafer. It would use very little of the precious material. Now, imagine if it was deposited on a sheet of film or thin metal or glass, without the time-consuming and costly work of slicing wafers from ingots with special equipment. Imagine the individual cells deposited next to each other, instead of being mechanically assembled. That's the idea behind thin-film technology. It is also called amorphous, meaning "not crystalline." The active material may be silicon, or it may be a more exotic material such as cadmium telluride.

These thin-film panels can be made flexible and lightweight by using plastic glazing. Some flexible panels can tolerate a bullet hole without failing. Some perform slightly better than crystalline modules under low-light conditions, such as when the sky is slightly overcast with a thin layer of clouds. Many websites still claim that thin-film PV panels are less susceptible to losing power on the entire array when only one or two modules are shaded. However, these early claims of superior shade tolerance seem to have been exaggerated.

PV Panel Characteristics Comparison: Crystalline vs. Thin-Film
CRYSTALLINE
Very low shadow tolerance.
100-watts peak crystalline panel is 25 percent smaller but produces 35 percent less power per year during the hot summer or in tropical climates.
Crystalline panels suffer a slightly higher voltage drop when heated by the sun in very hot climates, though modern crystalline panels now have three substrings, which has reduced output losses. Crystalline panels that are 16–17 percent efficient at 70–80°F will be only 10–12 percent efficient with a surface temperature of over 190°F.
Crystalline panels suffer from an average deterioration of 1.0 percent to 1.4 percent per year from the stated watts peak value in tropical climates. The original rated output is given as the output before any deterioration. Crystalline panels are much more efficient than thin-film, and they're considered more stable and reliable long-term. This is one of the reasons why crystalline panels have generally had longer warranty periods, compared to thin-film.
Crystalline cells are a bit more expensive than thin-film, unless you also consider space efficiency.
Crystalline panels are very expensive to make in smaller sizes, and this increase in cost prevents them from taking advantage of the smaller modular formats available with thin-film.

Conclusion:
Crystalline silicon PV panels have advantages that outweigh thin-film technology. In almost all cases, for homeowners in the US looking for a solar PV rooftop system, either monocrystalline silicon or polycrystalline silicon is recommended.

The major disadvantages of thin-film technology are lower efficiency and uncertain durability. Depending on the technology, thin-film module prototypeshave reached efficiencies between 7 and 12 percent, and production modules typically operate at about 9 to 10 percent. Some predict that future module efficiencies will climb to between 11 and 13 percent. Lower efficiency means that more roof space and mounting hardware are required compared to monocrystalline or polycrystalline modules to produce the same power output. Thin-film materials also tend to be less stable than those used in crystalline panels, and thin-film can suffer relatively more efficiency degradation during theinitial months of operation and beyond. However, thin-film technology is the subject of constant research. We will likely see many new thin-film products introduced in the coming years with higher efficiencies and longer warranties.

THIN-FILM

Slightly better shadow tolerance, though not as good as has been claimed by manufacturers. Even inrecent years, this was generally considered one of thin-film's biggest advantages, but in practice the performance of thin-film under slightly shady conditions is not noticeably better than crystalline silicon.

One-hundred-watts peak thin-film panel is 25 percent bigger but produces 10 percent to 15 percent more energy per year in hot climates. In these situations, thin-film gives slightly more energy for thesame surface area.

Thin-film panels perform at a slightly higher voltage output under direct heat and can still produceover 20 volts at 150°F or higher surface temperature. This would not apply in Canada, Europe, or most parts of the United States.

Over the long term, thin-film panels suffer from an average yearly deterioration of slightly less than1 percent per year from the stated watts peak value.
The thin-film panel rated output is given as the output after the initial 30 percent initial deterioration during the first 3 or 4 months. A thin-film panel will then have only slightly less deterioration than crystalline.

Thin-film cells are slightly less costly for a specified PV system output in kWh.

> The advantages of using small modular format solar voltaic thin-film panels include lower windresistance, fewer system failures (if a small module is broken, it makes little difference to the totaloutput), and fewer repair costs (replacing a small module is much cheaper than replacing a large broken panel).

> Conclusion:
> Only in rare cases and special circumstances should you consider thin-film.

The market for thin-film PV grew appreciably starting in 2002. In recent years, about 5 percent of US photovoltaic module shipments to the residential sector have been based on thin-film technology. It continues to have a sizeable market for small electronic devices and special applications, including building- integrated photovoltaics (BIPV). Thin-film technology has also been popular forvery large solar farms in deserts, where the land cost is very low and the solar insulation index is very high.

Two DIY installers holding a long, amorphous silicon thin-film panel that can easily adapt to a curved substrate or surface.

TYPES OF THIN-FILM SOLAR CELLS

Thin-film solar cells are manufactured by depositing one or several very thin layers of photovoltaic material onto a substrate. They're also known as "thin-film photovoltaic cells" (TFPV). The different types of thin-film solar cells are generally categorized by the kind of photovoltaic material that is deposited ontothe substrate. Commercial production of thin-film solar cells is currently basedon three basic types: amorphous silicon (A-Si), cadmium telluride (CdTe), andcopper indium gallium selenide (CIGS). Very recently, a fourth type has been developed, namely organic solar cells. Solar panels based on amorphous silicon, cadmium telluride, and copper indium gallium selenide are currently the only thin-film technologies commercially available on the market. Organic or polymer solar cells may be available for some applications by 2020.

A fairly common ground-mounted thin-film PV solar farm.

An organic solar cell or plastic solar cell is a type of photovoltaic that uses organic electronics, a branch of electronics that deals with conductive organic polymers. An example of an organic photovoltaic is the polymer solar cell. As of late 2015, polymer solar cells were reported in research projects to have reached efficiencies of up to 10 percent. However, figures for production models will notbe available for several years. These PV cells have low production costs and may be cost-effective for some photovoltaic applications. The main disadvantages associated with organic photovoltaic cells are low efficiency, low stability, and low strength compared to crystalline silicon solar cells. This type of organic solar cell is not yet considered commercially competitive, and the technology is still in the early stages of research and development. Therefore, thehomeowner looking for a solar PV rooftop system should not seriously consider it.

Because their electrical power output is low, solar cells based on amorphous silicon have traditionally only been used for small-scale applications such as in- pocket or portable calculators. However, recent innovations have made them more attractive for larger scale applications. With a manufacturing technique called "stacking," several layers of amorphous silicon solar cells can be combined, resulting in higher efficiency rates (in the 6 to 8 percent range). Onlyabout 2 percent of the silicon used in crystalline silicon solar cells is required in the manufacture of amorphous silicon solar cells. This makes a-Si cells extremely cost competitive per watt if you ignore the cost and availability of thespace required. On the other hand, stacking is a relatively expensive process.

Cadmium Telluride (CdTe) Solar Cells

Cadmium telluride (CdTe) photovoltaics make use of a photovoltaic technology that is based on the use of cadmium telluride thin-film, a semiconductor layer designed to absorb and convert sunlight into electricity. Cadmium telluride is theonly thin-film solar panel technology that, in some cases, has surpassed the cost-efficiency of crystalline silicon solar panels for multi-megawatt systems. The average efficiency of solar panels made with cadmium telluride is generally considered to be in the range of 10 to 12 percent.

First Solar has installed over five gigawatts (GW) of cadmium telluride thin-film solar panels worldwide. First Solar holds the world record for CdTe PV module efficiency at 14.4 percent.

Flexible thin-film solar cells that can be produced by roll-to-roll manufacturing are a highly promising route to cheaper solar electricity. Scientists from Empa, the Swiss Federal Laboratories for Materials Science and Technology, have made significant progress in paving the way for the industrialization of flexible, lightweight, and low-cost cadmium telluride (CdTe) solar cells on metal foils. They also succeeded in increasing the efficiency rating of the cells from below 8 percent up to 11.5 percent by "doping" the cells withcopper.

After crystalline silicon, CdTe solar cells are the next most abundant photovoltaic product in the world, currently representing about 4 percent of theworld market. CdTe thin-film solar cells can be manufactured quickly and inexpensively, providing a lower-cost alternative to conventional silicon-based technologies. The record efficiency for a laboratory CdTe solar cell is 18.7 percent, which is well above the efficiency of current commercial CdTe modulesthat run between 10 percent and 13 percent.

Polyimide film is a new material currently in development for use as a flexible superstrate for cadmium telluride (CdTe) thin-film photovoltaic modules. Because Kapton film is over 100 times thinner and 200 times lighter than the glass typically used for PV panels, there are inherent advantages in transitioning to flexible, film-based CdTe systems. High-speed and low-cost roll-to-roll adhesion technologies can be applied for high-throughput manufacturing of flexible solar cells on polymer film. The new polyimide film potentially enables significantly thinner and lighter-weight flexible modules that are easier to handle and less expensive to install, which would make them a candidate for many applications, including building-integrated photovoltaics.

However, due to low efficiencies and other limitations, as well as the premature stage of development of polyimide film PV technology, this type is not recommend to the homeowner looking for a PV solar rooftop system.

As it provides a solution to key issues like climate change and water scarcity, CdTe PV is considered the most eco-friendly technology among the available types, and it may provide some element of energy security. It's also considered the most eco-efficient PV technology when comparing a range of applications, including installation on commercial rooftops or large-scale ground-mounted PV systems. Some claim that CdTe PV has the smallest carbon footprint, lowest water use, and fastest energy payback time of all solar technologies.

CIGS component elements are used to make thin-film solar cells (TFSC). Compared to the other thin-film technologies summarized above, CIGS solar cells have demonstrated the most potential in terms of efficiency. These solar cells contain lower amounts of the toxic material cadmium than is found in CdTe solar cells. Commercial production of flexible CIGS solar panels was initiated in Germany in 2011. The efficiency ratings for CIGS solar panels are in the range of 10 to 13 percent, but this may increase over time.

CIGS features much higher absorption than silicon, so a layer of CIGS can absorb more light than a silicon layer of the same thickness. With thin-film, some of the light-gathering efficiency is given up in exchange for the advantages of thinness. But with the highly absorptive CIGS, the efficiency trade-off is less severe than with silicon PV cells. The record efficiencies for thin-film CIGS cells are slightly lower than that of CIGS used in lab-scale, top-performance cells, which are rated at 19.9 percent efficiency. Compared with those achieved by other thin-film technologies such as cadmium telluride or amorphous silicon, this is the highest efficiency rating reported.

A ground-mounted solar array.

CIGS solar cells are not as efficient as crystalline silicon solar cells, for which the record efficiency is over 25 percent. However, many companies argue that CIGS is substantially cheaper due to lower fabrication costs and significantly lower material costs. A direct band gap material, CIGS has very strong light absorption. Two micrometers of CIGS is enough to absorb most of the sunlight that strikes it; a much greater thickness of crystalline silicon is required for the same amount of absorption.

The active layer of CIGS can be deposited directly onto molybdenum-coated glass sheets or steel bands. This takes less energy than growing large crystals, which is a necessary step in the manufacture of crystalline silicon solar cells.

Also, unlike crystalline silicon, these substrates can be flexible, a notable advantage when it comes to design and fitting solar cells to curved surfaces, as demonstrated in the photos below.

Many thin-film solar cell types are still early stages of research and testing. Some seem to offer promising potential, and we can expect to see more of them in the future. Indeed, the initial attraction of CIGS was its promise of lower-cost manufacturing, both in terms of the materials required and in its streamlined, roll-to-roll manufacturing process. However, crystalline silicon market prices have decreased so much in recent years that this former advantage of CIGS has been erased.

Three CIGS arrays of four panels, each suspended by rope or cable demonstrate their flexibility.

An attractive example of BIPV with the PV solar system installed on this very steep, two-roof house in Colorado where there is often big snowfalls in winter.

C-Si (crystalline silicon) manufacturing has become much more streamlined and standardized, while CIGS remains a customized technology. And CIGS's conversion efficiencies haven't kept pace with c-Si. As of recently, Miasole wasthe current CIGS module record-holder, at 15.7 percent efficiency.

Building-Integrated Photovoltaics (BIPV)

Lastly, I'll briefly touch on the subject of building-integrated photovoltaics. Rather than being an individual type of solar cell technology, building-integrated photovoltaics have different methods of integration. BIPV can be developed with either crystalline or thin-film solar cells. They can be incorporated into facades, roofs, skylights, windows, walls, and other building surfaces that can bemanufactured with one of the basic photovoltaic cell materials listed above.

Smaller surfaces like covered walkways and solariums can also be designed andbuilt as mini-BIPV projects.

If you have the extra money and want to seamlessly integrate photovoltaics with the rest of your home or condominium project, you might consider building-integrated photovoltaics, especially if your home is still in the designstage. These panels are extremely versatile and can be used to replace conventional building materials for practically any part of the building's exterior. BIPV is increasingly being incorporated into the construction of newbuildings as a principal or back-up source of electrical power, although existing buildings may also be retrofitted with similar technology. The advantage of building-integrated photovoltaics over more common nonintegrated systems is that the initial capital cost can be substantially offset by reducing the amount spent on building materials and labor that would normally be used to construct the roof or the other part of the building that the BIPV modules replace. These advantages make BIPV one of the fastest-growing segments of the photovoltaic industry for residential, commercial, and industrial applications.

How to Choose the Best Solar Panel Type for Your Home

After you've done some basic research by reading this book, and after you have a pretty good outline of your PV solar rooftop project, it's a good idea to have your project evaluated by an expert. This will help you determine what type and quantity of solar panels would be best for your home.

One aspect to consider is space limitations. The majority of homeowners don't have enough space for thin-film solar panels. In this case, crystalline silicon-based solar panels are usually your best choice. Regardless of space issues, though, you'll likely want to choose between monocrystalline and polycrystalline PV modules. Furthermore, in many areas there are no residential solar installers who offer a thin-film solar panel option.

Summary of Advantages and Disadvantages of Basic Thin-Film Technologies			
TECHNOLOGY	**MAXIMUM DEMONSTRATED EFFICIENCY**	**ADVANTAGES**	**DISADVANTAGES**
Amorphous Silicon (a-Si)	12.2 percent	Mature manufacturing technology	Low efficiency; High equipment costs to manufacture
Cadmium Telluride (CdTe)	16.5 percent	Low-cost manufacturing	Medium efficiency; Rigid glass substrate
Copper Indium Gallium Selenide (CIGS)	19.9 percent	High efficiency; Glass or flexible substrates	Film uniformity is a challenge on large substrates
Organic Solar Cells or Polymer Solar Cells	10 percent	Low manufacturing cost; Wide variety of possible substrates; Very light weight	Instability problems; Efficiency problems; Still in early stages of R&D as of 2016

You'll likely have a choice of different solar panel sizes. The 180, 200, and 220-watt rated solar panels (and even higher wattages) are usually the same physical size. They're manufactured in exactly the same way but perform differently when tested, and hence they fall into different categories for power output. If available area is very limited, you'll logically select the highest-rated power output for a particular physical size of module, and that will be monocrystalline.

Both monocrystalline and polycrystalline solar panels are good choices and offer similar advantages. Even though polycrystalline solar panels tend to be less space efficient and monocrystalline solar panels tend to produce more electrical power, this is not always the case. It would be nearly impossible to recommend one or the other without examining the alternative solar panels available, as well as your specific location and its physical conditions.

Monocrystalline solar panels are slightly more expensive but also slightly more space efficient. If you had one polycrystalline and one monocrystalline solar panel, both rated 220-watt, they would generate the same amount of electricity. But the one made of monocrystalline silicon would take up slightly less space, and this can be an important factor in deciding which type of PV module to use.

Two large arrays of poly panels mounted on the sloping roof of a low-rise commercial building.

In addition to the different PV technology descriptions and comparisons, there are numerous other factors and considerations you should take into account before making a final decision about which solar panel to buy.

For homes and buildings with ample roof space, a panel's peak efficiency shouldn't be the primary consideration. It's more important to consider the system as a whole, balancing price with quality. Where space is really limited, efficiency considerations may outweigh the desired output of the solar system. These, of course, will carry a higher price tag. But for the budget-conscious homeowner, the number to look at will be dollars-per-watt for the entire PV system.

A small array of mono panels mounted flush to the surface of a wooden shingle roof. Wood roof shingles are normally made of cedar that has a natural attractive finish and they have a long useful life to match the expected useful life of PV panels.

A close-up shot of the mid-bracket securing two poly panels to the support railing below (not visible here).

In the end, the cost and performance of your system will depend not only on the panels you use, but also on the inverter you choose. If you hire a contractor, his or her installation costs will also come into play. Performance will also be affected by the east-west orientation (azimuth) of your roof and the tilt angle of your panels, which are commonly installed at an angle parallel to the roof.

It's also important to look beyond module-efficiency ratings. As a prospective solar system owner, you'll need to consider the company behind the product you're buying. Although quality technology is important in the selection of solar panels, you must remember that both monocrystalline and polycrystalline silicon solar cells are proven technologies, and one should not automatically be considered better than the other.

Manufacturing equipment for silicon wafers is more readily available than ever. The market for PV solar panels is relatively easy for large companies to enter. A critical difference between quality manufacturers and others is whether the company in question invests seriously in research and development. R&D investment by a manufacturer is indicative of the company's commitment to creating innovative and high-quality products. It also shows that the company is planning to be active in the market for a long time.

A small array of six poly panels on a small steep roof.

For most households, balancing affordability with reliability is important.

Solar-power systems are expected to work for 30 years or more, and warranties are usually valid for 25 years. But warranties are only useful if the manufacturer remains a solvent company. Though it's impossible to know what will happen to a company 15 or 25 years from now, it's still wise to form an idea whether the manufacturer is financially sound and likely to be around for at least as long as the warranty period. If panels need repairing or replacing, the cost could become substantial if the warranty is of no use. Most of the PV module manufacturers are public companies. This makes it easy for you to ascertain if they're financially solid and if they are seriously investing in R&D. These criteria will determine if the warranty is of real value.

A DIY installer inspecting his mono panels up on an aluminum ladder. This kind of ladder is not recommended for use when wiring the PV rooftop system as it can conduct electric current. I always insist that DIY PV system installers use only wooden ladders.

Chapter 2 - Types of PV System Connections

Once you have a general understanding of how the solar-energy system functions, choosing the right parts will be easier. It's not complicated. I'll now describe the parts of a standard grid tie, or "on-grid PV system." We'll also look briefly at what is known as an "off-grid PV system," and I'll explain the differences between the on-grid and off-grid systems and the normal conditions and criteria for selecting a system.

First, let's look at the components of the grid-tied or on-grid system so that you know what you need to buy and what to plan for. Then I'll review some of the options available for the various component parts. I'll show you how to determine the size of the system you want and the positioning and layout of the panels, as well as the factors you'll want to consider for the mounting framework that will attach the panels to your roof. We'll also discuss some installation details that will complement the step-by-step photographic installation procedures illustrated and explained in Chapter 10. Lastly, I'll review some financial assistance possibilities that may be available to you.

Basically, an on-grid PV system involves staying connected to the power grid. This means you'll still receive power from the utility company when you need it, and in many areas you'll be able to sell your excess power back to the utility under a simple registered agreement. Selling the excess power back will involve the installation of a net metering system that tracks and records how much credit you receive for the excess power your system feeds back into the grid.

An off-grid PV system is a solar electric generating system that is usually relatively small and is not connected to the utility grid. There can be many reasons for installing an off-grid PV system. It might be that the location of the system is not anywhere close to the nearest grid connection. The off-grid system might be used for a remote cabin, a small ranch, a hunting lodge, or for a remote research station or microwave station. Usually an off-grid PV system will include a battery storage subsystem, as there will be no power available from the grid for energy requirements after dark or in bad weather.

What are the benefits of grid-connected solar panels versus living off the grid? Deciding whether or not to grid-tie your solar array is usually pretty straightforward. The clear-cut benefits of being grid-tied appeal to most homeowners. There are, however, some who choose to live off-grid.

Let's take a closer look at the benefits and downsides of grid-tied, off-grid, and hybrid connection solar systems.

Grid-Tied Solar Systems

Grid-tied, on-grid, utility-interactive, grid inter-tie, and grid back-feeding are allterms used to describe the same concept: a solar PV system that is connected to the utility power grid.

One advantage of grid-tying is that compared to the other two basic connection types, a grid connection will allow you to save more money with your solar PV array through lower rates granted by the utility company, net metering, and lower equipment and installation costs. For a fully functional off-grid solar system, batteries and other stand-alone equipment are required, andthese will add to both the initial cost of the system as well as maintenance costsdown the road. Grid-tied solar systems are therefore generally cheaper and simpler to install, as long as the grid is reasonably close to your home.

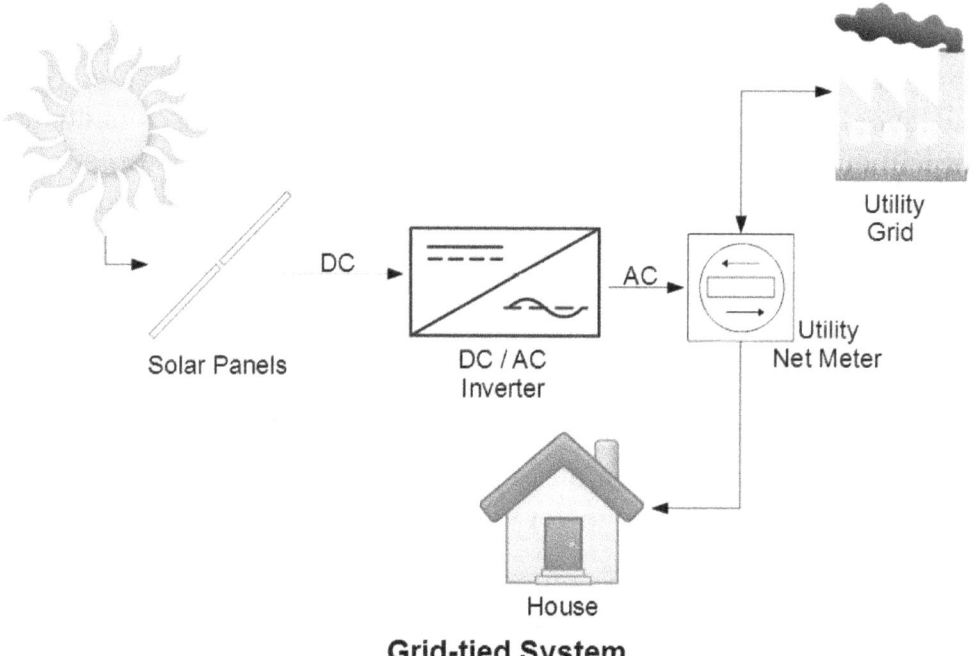

A schematic of a "grid-tied" solar system.

On bright days, your solar panels will usually generate more electricity than you will consume during the daylight hours. With net metering, homeowners canput this excess electricity onto the utility grid rather than store it themselves in a battery-bank storage system, which would involve a considerably larger initial investment.

Net metering (called a "feed-in tariff" scheme in many countries) has playeda very important role in how solar power has been supported by governments toincentivize new buyers. Without it, residential solar systems are much less financially feasible. With net metering, most utility companies are required to buy electricity from homeowners at the same rate as they sell it themselves. Inthis way, the utility grid acts as a huge virtual battery system.

Let me explain this battery concept. Electricity has to be spent in real time. However, it can be temporarily stored as other forms of energy (e.g. chemical energy in batteries), though storage involves significant losses. The electric power grid also works like a battery, but without the need for maintenance or replacements and with much higher efficiency. In other words, more electricity and more money goes to waste with conventional battery systems compared to grid-tied systems.

Annual US electricity transmission and distribution losses average about 7 percent of the electricity transmitted, according to the Energy Information Administration (EIA) data. Lead-acid batteries, which are commonly used with solar panels, are only about 80-85 percent efficient at storing energy, and their performance degrades over time depending on factors like rate of discharge and the degree of discharge, or "DOD." The DOD is the minimum charge level at which point your PV system stops draining the battery and starts the recharge cycle again. The battery "charge controller" manages this DOD cut-off level.

This DOD control system is used to protect the battery or battery bank because if you drain the batteries too much it will reduce their useful life.

Additional perks of being grid-tied include access to backup power from the utility grid in case your solar system stops generating electricity for one reason or another. At the same time, you'll also help to mitigate the utility company's peak load. As a result, the overall efficiency of the electrical grid system is improved.

Equipment Components Of A Grid-Tied SolarSystem

There are a few key differences between the equipment needed for grid-tied, off-grid, and hybrid PV solar systems. In addition to the PV panels, mounting support system hardware, and other standard items like DC and AC disconnect switches, grid-tied solar systems rely on grid-tie inverters (GTI) or micro- inverters, and power meters (net metering type).

Grid-Tied Inverters (Gti)

What's the job of PV solar system inverters? They regulate the voltage and convert the direct current (DC) received from your solar panels. The direct current generated by your solar panel array is converted by a central inverter or by many micro-inverters into alternating current (AC), which is the type of current utilized by the electrical appliances in your home. It might be noted here that it's now possible in some areas to purchase DC appliances. In addition, grid-tied inverters, also known as grid-interactive or synchronous inverters, synchronize the phase and frequency of the current to fit the utility grid, nominally 60Hz. The output voltage is also adjusted slightly higher than the grid voltage, in order for excess electricity to flow outwards to the grid.

Power Meters (Net Metering Type)

Most homeowners will need to replace their current power meter with one designed for net metering. This device, often called a net meter or a two-way meter, is capable of measuring power going in both directions, from the grid to your house and vice versa. You should consult with your local utility company and see what net metering options you have. In some places, the utility company will issue a power meter for free, and they'll pay full price for the electricity you generate and deliver to the grid (the same price $/kWh as they charge you).

However, this is not always the case.

Off-Grid Solar Systems

An off-grid solar system is the obvious alternative to a grid-tied system. For homeowners with access to the grid, off-grid solar systems are usually not preferred because in order to ensure access to electricity at all times, off-grid solar systems require battery storage and a backup generator. On top of this, a battery bank typically needs to be replaced after ten years. Batteries are complex, they involve a high initial cost with respect to the overall system, and they also slightly decrease overall system efficiency. However, an off-grid system can provide a viable solution in the case that you have no access (or very difficult access) to the utility grid. This type of system can be cheaper than extending power lines in most remote areas, so you might consider an off-grid system if your location is more than 200 yards from the grid. The cost for new overhead transmission lines start at around $174,000 per mile ($108,000 per Km) for rural construction, and they can run into the millions of dollars per mile for urban construction.

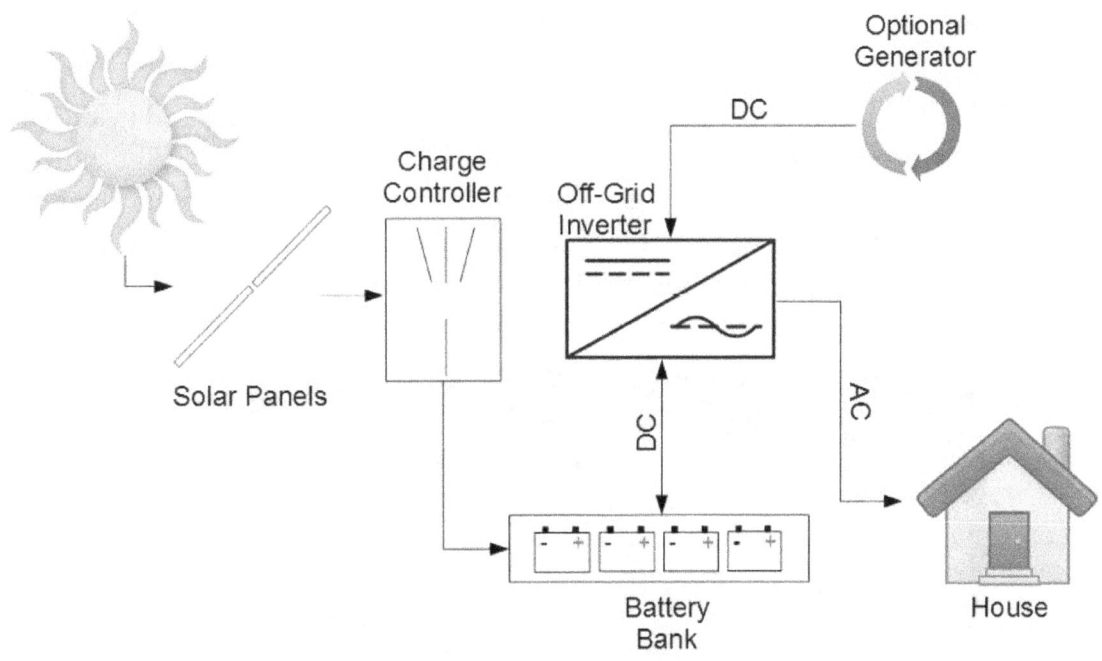

Off-Grid System

A schematic of an "off-grid" solar system.

Off-grid systems also provide a solution for those who want to be energy independent. To many, the idea of living off the grid and being self-sufficient just feels good. For some people, this feeling is worth more than saving money. Energy self-sufficiency is also a form of security. Local or regional power failures in the utility grid do not affect off-grid PV solar systems. On the other hand, batteries can only store a limited amount of energy, and during cloudy times, sometimes security actually lies in being connected to the grid. To manage this risk, you might be wise to consider installing a backup diesel-fuel electric generator to be prepared for extended periods of rainy and cloudy weather.

Typical off-grid solar systems require the following extra components in addition to the PV solar modules and mounting support framework hardware:

- Solar Charge Controller
- Battery Bank
- DC and AC Disconnect Switches
- Off-Grid Central Inverter or Micro-Inverters

- Backup Generator (optional but recommended)

Solar charge controllers are also known as charge regulators or battery regulators. The last term is probably the best for describing what this device actually does: Solar battery charger controllers limit the rate of current being delivered to the battery storage bank and protect the batteries from overcharging. Good charge controllers are crucial for keeping batteries healthy and ensuring that the lifetime of a battery bank is maximized. If you have a battery-based inverter, chances are the charge controller is integrated into the inverter.

A "battery bank" of 10 batteries.

Battery Banks

Without a battery bank or a backup generator, an off-grid system will mean "lights out" at sunset. A battery bank is essentially a group of batteries wired together, connected to the inverter with a DC disconnect switch, and then connected to selected wire circuits to provide power after dark to certain lights, as well as to selected electrical outlets for appliances like a television, computer, or refrigerator. LED lights will help extend the power provided before the battery bank runs out of energy.

The above photos illustrate the diversity and flexibility of PV solar batterybank systems:

A DC disconnect switch, a meter, and an AC disconnect switch.

A different DC disconnect switch.

The inside view of a DC disconnect switch.

The beauty of solar battery banks is they're completely modular, which enables the homeowner to expand the energy storage system whenever desired. Depending on proper maintenance, frequency of use, and "depth of discharge," the solar batteries will be serviceable for up to ten years before they'll need to be replaced. We go into more details about battery systems and back-up generatorsin Chapter 3.

DC Disconnect Switch

AC and DC safety disconnects are required for all solar systems. For off-grid solar systems, one additional DC disconnect is installed between the battery bank and the off-grid inverter. It's used to switch off the current flowing between these components. This is important for maintenance, troubleshooting,and protection against electrical fires.

Off-Grid Inverters

There's no need for an inverter if you're only setting up solar panels for your boat or your RV, or for electric loads that use only direct current. However, you'll need an inverter to convert DC to AC for all other standard electrical appliances in your home. Off-grid inverters do not have to match phase with theutility sine wave as opposed to grid-tied inverters. Therefore, you can expect agiven manufacturer's off-grid converter to be technically simpler and slightly more economical than its grid-tied counterpart. Electrical current flows from thesolar panels through the solar charge controller and the battery bank before it's finally converted into AC by the off-grid inverter.

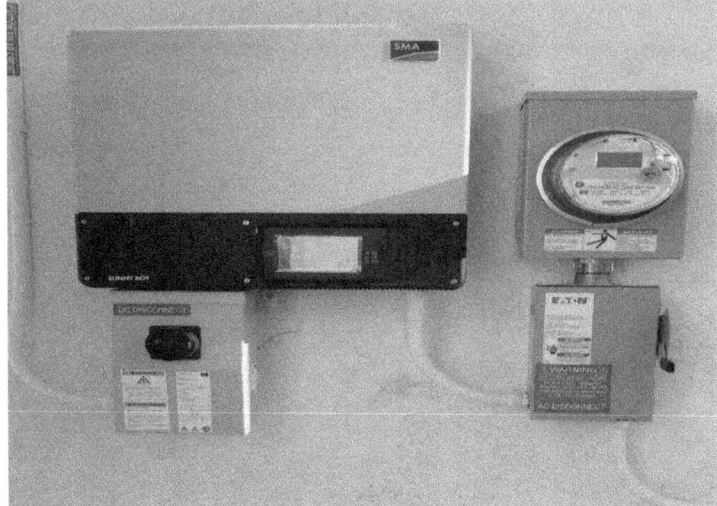

A metering device with a DC disconnect switch.

It takes a lot of money and big batteries to prepare for several consecutive days without the sun shining or access to the grid. This is where backup diesel generators can play a key role. In most cases, installing a backup generator that runs on diesel is a better choice than investing in an oversized battery bank that seldom gets to operate at full potential. Backup generators will normally run on propane, petroleum (diesel), or gasoline. Backup generators typically produce an AC output, which can be sent through the inverter for direct use or can beconverted into DC for battery storage.

Hybrid Solar Systems

Hybrid solar systems combine the best elements from both grid-tied and off-gridsolar systems. These systems can either be described as off-grid solar with utilitybackup power, or grid-tied solar with backup battery storage. If you own a grid- tied solar system and drive a vehicle that runs on electricity, you already have something similar to a hybrid system. In essence, electric vehicles are battery systems on four wheels.

A hybrid solar system is a grid-tied system with backup energy storage that incorporates a battery bank. Hybrid solar systems are less expensive than off-grid solar systems. Often you don't really need a backup generator, and the capacity of your battery bank can be downsized. Off-peak electricity from theutility company is cheaper than using a backup diesel generator. Also, the noiseof a generator motor may be an issue to consider.

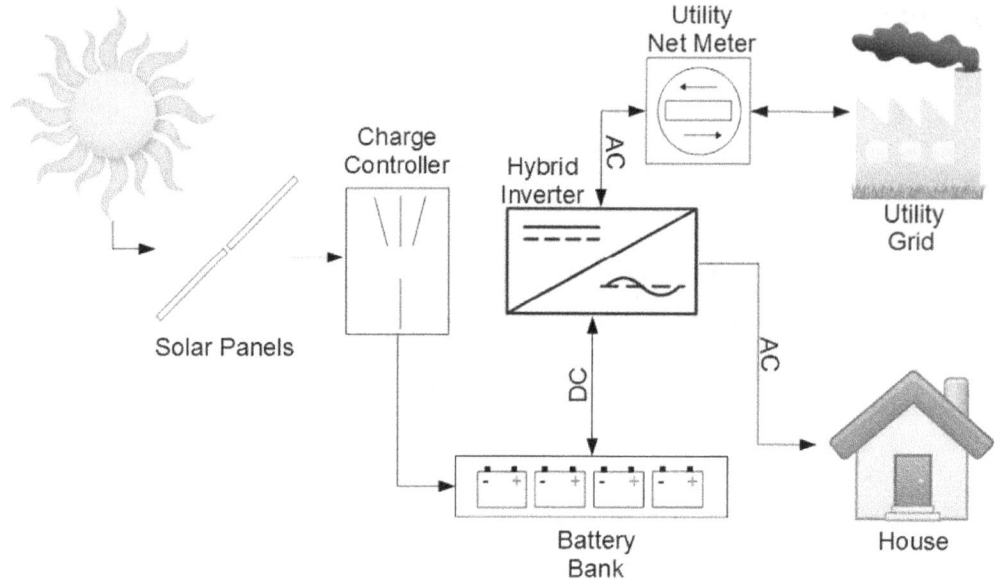

Hybrid System

A typical hybrid connection solar system often simply referred to as a "hybrid solar system."

Hybrid solar systems utilize battery-based grid-tie inverters. Combining these devices can draw electrical power to and from battery banks, as well as synchronize with the utility grid. The new inverters let homeowners take advantage of changes in the utility electricity rates throughout the day. Solar panels usually output the most electrical power around noon, just before the time when the price of electricity peaks. Your home can be programmed to consume power during off-peak hours by appropriately programming your rooftop PV solar system. Consequently, you can temporarily store excess electricity from your PV solar panels in your battery storage system, and then automatically feed it into the utility grid at a time when you will be paid the most for every kWh.

Smart solar holds a lot of promise, as the introduction of hybrid solar systems has opened up opportunities for many interesting innovations. The concept will become increasingly important as we transition towards the so-called "smart grid" in the coming years.

But for now, the vast majority of homeowners who have a solar energy system find that tapping the utility grid for electricity while also using the grid for energy storage is significantly cheaper and more practical than using hybrid systems with battery banks and backup generators. Complete independence from the grid system may be attractive to many on principle, but it increases the cost of the initial investment substantially. Therefore, this option is usually discarded in favor of a simple grid-tied solar system.

Of course, where the home or weekend cottage is in a remote location without access to the grid or where the cost of installing a line is prohibitive, owners might gladly pay for the extra cost of a battery storage system, and maybe also a generator.

Chapter 3 - Major Components of a Solar PV Rooftop System

When choosing a solar energy system for your home, there are three major components that you'll need to be concerned about: solar modules, solar racking or supportframes, and inverters. The components you choose will determine the reliability and output of your solar array for the duration of the system's time on your home.

Solar PV Modules and Mounting Support Hardware

In a grid-tied system, electricity is initially generated by several to many PV solar modules. Chapter 2 provided a detailed description and analysis of the various types of solar modules available. Once the type and size of the systemare chosen, a simple rooftop mounting frame, to which the modules can be attached, will be installed. See example in the photo on the following page.

There are some good options for solar mounting equipment. Solar trackingdevices are now more economical and easier to install than earlier systems, withnew products coming to market all the time.

Of course, the mounting frame is installed on the side of the roof that provides maximum sunlight exposure. In the northern hemisphere, this would logically be on the south-facing part of the roof, which is slanted toward the equator. The more perpendicular to the equator the mounting frame supports are,the better, but this line does not need to be exactly 90 degrees for the PV solar system to function well.

A top view of three rows of support rails mounted on the roof with micro-inverters already attached and waiting to be connected to the PV modules.

The type of mounting system you'll need depends on where you plan to install the modules of your PV solar energy system. With crystalline panels, you have three options for location: mounting the panels on your roof, on the ground, or on a pivoting stand.

Most people think a roof mount is the most convenient and aesthetically pleasing, but there are many reasons that people choose other options. For example, if your roof is small, unstable, in the shade, or if you aren't able to face the panel towards the equator (facing south in the northern hemisphere or facing north in the southern hemisphere), you may consider mounting your system elsewhere. You may find that you like the simplicity of a ground mount if you have extra land available.

Pivoting stands are an attractive alternative because they're able to follow the sun throughout the day, so they can be far more efficient. But they're also more expensive. If you have enough open space and your roof has major disadvantages, then a ground mount may be the best choice. With any of these mounting options, you should make sure there are no local ordinances or homeowners association rules against them.

There are viable solar panel racks and mounting hardware for every solar panel installation style; your solar array depends on your property. If you're not sure about the different choices, any qualified solar contractor can help you make the right decision.

Two rows of support rails installed and waiting for installation of the PV panels and, if applicable, the micro-inverters first go on the rails before being connected to the PV panels.

Flush support frames are the most popular choice. They provide an inexpensive and simple option that is suitable for most roof-mounted solar panel installations. They're generally not adjustable and are usually designed to position the solar panels at a consistent height above the surface of the roof on which they are mounted. There are suitable leak-proof anchor bolts or fasteners for any kind of roof. Tilted mounts, which improve the angle of the PV panels for roof installations, are also available. These are definitely recommended for flat roofs.

Ground-mounted support frames can provide an excellent alternative—space permitting, and assuming there are no serious shading problems—that will allow you to install your solar panels at ground level anywhere on your property.

Ground mounts are usually designed with a fixed-tilt angle. But they're often adjustable, allowing you to tilt your solar modules to the appropriate level for optimal solar collection, depending on your latitude.

There are also fully automated, adjustable-tilt tracking mechanisms that can substantially increase the solar efficiency of your system. Generally speaking, the lower the latitude, the less tilt your mount will require. Four-way tilt-control tracking systems can maintain the panels in a perpendicular position to the sun's rays from east to west, following the arc of the sun.

For those requiring a non-roof installation, pole mounts are another solution. These mounts come in one of three alternative types that are distinguished from one another by how they're positioned on the pole: top of pole, side of pole, and pole with tracking device.

Regardless of the type of mounting you and your solar contractor decide on, your support frame or pole-mount system will provide safety and security for your solar investment.

While most people might want to focus on panels and inverters, it's important to remember that the solar panel mounting frames can also be critical to the success of your overall system. Besides the orientation and shading issues we discussed above, you also need to find out the wind category for your area, and also check the conditions of the ground if you plan to install a good ground-mount system.

Even with off-the-shelf parts, many permit offices will not give you a permit if the proposed ground-mount system doesn't have a civil engineer's stamp of approval. You should be able to hire an engineer for $500 or even less, depending on your location.

Dc And Ac Disconnect Switches

Now let's look at another component of a PV system: disconnect switches. A shutoff switch, more commonly known as a "disconnect switch," separates the panels from the rest of the system so that you'll be safe if you ever need to make any repairs or changes to the system, such as adding additional panels. The first disconnect switch is the DC disconnect switch.

The DC disconnect switch is relatively small and can be easily installed anywhere convenient along the chosen routing for the PV cables that connect the rooftop modules to the inverter. Usually, you would choose your routing for the PV cables in the least conspicuous layout, taking advantage of any chimney, rooftop sun windows, attic bedroom roof protrusions, or eave troughs. You will want to pick a color for your PV cables that will tend to blend in with the existing rooftop colors, most often dark brown. Here are a few photographs of standard disconnect switches.

A DC disconnect switch, central inverter, AC disconnect switch, and metering device.

An off-grid disconnect switch.

A solar PV system typically has two safety disconnects. The first is the PV disconnect (or Array DC Disconnect). The PV disconnect allows the direct current coming from the modules to be interrupted before reaching the inverter for service and safety reasons.

The second disconnect is the AC disconnect. The AC disconnect is used to separate the inverter from the house's main distribution panel and also from the electrical grid. In a solar PV system, the AC disconnect is usually mounted to the wall between the inverter and utility meter. The AC disconnect may be a breaker on a service panel, or it may be a stand-alone switch. This disconnect is sized based on the output current of the inverter or micro-inverters.

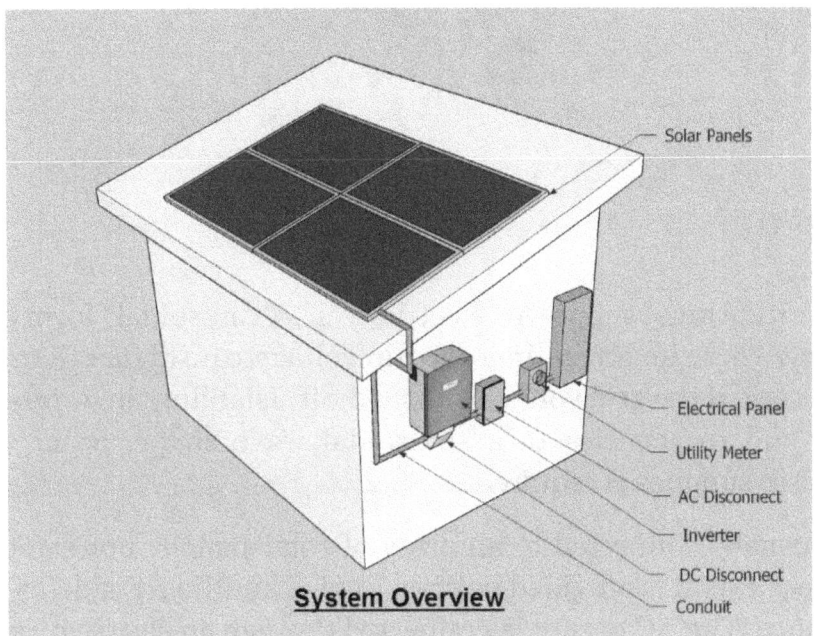

A schematic of a simple, small PV system installed on a garage.

Solar Pv System Grid-Tied Inverters

Having chosen your solar modules, you'll need to decide on a solar inverter.

Solar inverters are a critical component to your PV solar energy system. Inverters change the direct current coming from the panels into the alternating current that your appliances and light fixtures use. There are several considerations here. First, you need an inverter that can handle what your panels generate, so make sure the wattage of the inverter is at least as strong as the wattage of your array. Second, you can consider solar micro-inverters, which are smaller inverters that connect to each panel instead of a central inverter that converts the DC of the entire system as a whole.

Below is a photograph of a typical inverter unit. There are several well-known manufacturers of inverters, including SMA and InfluxGreen. They all make quality equipment that has been tested and approved by national and international electric standards authorities.

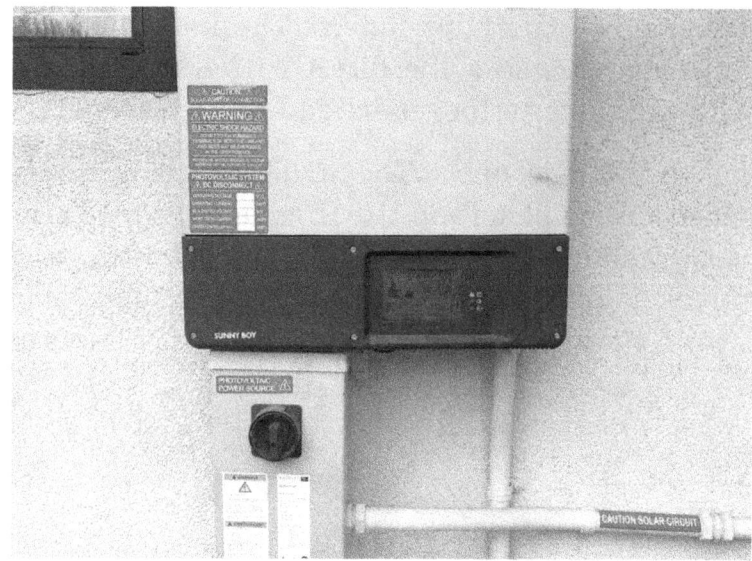

A typical central inverter.

Grid-connected inverters must supply AC electricity in a "sinusoidal" form that is synchronized to the grid frequency. These inverters limit feed-in voltageto no higher than the grid voltage and disconnect from the grid if the grid voltageis turned off. Islanding inverters need only produce regulated voltages and frequencies in a sinusoidal wave shape, as no synchronization or coordination with grid supplies is required.

A solar inverter is usually connected to an array of solar panels. For safety reasons, a circuit breaker or disconnect switch is provided both on the DC andthe AC side of the central inverter to enable maintenance. The AC output is connected through an electricity meter into the public grid and into the home'smain panel. In some installations, a solar micro-inverter is connected to eachsolar panel. We'll discuss the subject of micro-inverters in more detail a little later.

Off-Grid Inverters

Single phase off-grid PV inverters such as those made by InfluxGreen (Models IGSCI-0.5kVA to 7.0kVA) are an integrated system comprising a solar charger, AC charger, inverter, AC bypass switch, and transformer—all with battery control options. This inverter is versatile and can be used with a grid-tied systemor with an off-grid system where the grid is not readily available. The IGSCI solution provides a cost-effective, reliable, and efficient stand-alone system with battery backup to meet specific user needs.

Solar installations typically have a central inverter that takes the direct current generated by a group of PV panels and converts it into AC for the home and/or for the electrical grid. Micro-inverters function essentially the same way but are installed on the back of each solar panel. They have some distinct advantages over central inverters. Micro-inverters make a rooftop PV system or large commercial PV installation more efficient. They also make it easier to monitor power generation of each panel and quickly pinpoint a failure in the system. This can save a lot of time and money. The idea is basically that the added expense of using micro-inverters is more than offset by the savings created by increased efficiency and lower maintenance costs. Hanwha SolarOne is one of several manufacturers that incorporate a micro-inverter into every solarmodule.

A standard Enphase M215 micro-inverter—industry leader for many years.

A side view of the same three rows of support rails mounted on the roof with micro-inverters already attached.

When using a central inverter, economies of scale can certainly decrease thecost-per-kilowatt-hour of your solar PV system as you increase the size of the array. Many DC/AC central inverters are sized for systems up to five kilowatts. However, if your PV array is smaller (three or four kilowatts, for example), you might still be wise to buy the same 5 kW inverter. This is because the differencein the cost of the inverters is minimal. But this will facilitate adding more PV panels later on and will end up saving you money.

Until recently, central inverters dominated the solar industry. The introduction of micro-inverters marks one of the biggest technology shifts in thePV industry to date. Manufacturers of micro-inverters claim at least a 15 percentincrease in power output over central inverters, which in the long run can result in substantial savings for the homeowners who choose micro-inverters for their PV rooftop system.

Micro-inverters have been available since 1993. In 2007, Enphase Energy was the first company to build a commercially successful micro-inverter. Morethan one million units of the Enphase Micro-Inverter Model M175 have been sold since its release in 2008. Several other solar companies have since followed suit and launched their own micro-inverters, validating their potential andreliability.

Nine poly PV modules mounted after having the micro-inverters connected from the support rails and then to the modules.

The conduit wiring from the micro-inverters across the roof and down to the electrical control equipment.

Micro-inverters are quite small and are installed on the back of each solar panel in the upper right-hand corner. Alternatively, the micro-inverters may come separately and installed directly onto the support rails before being connected to the PV modules. Each micro-inverter is protected by the aluminum edge of the PV modules. In this way, the inverter doesn't interfere with stacking the panels when they're laid flat on top of one another for shipping and storage.

There has been a lot of debate on whether micro-inverters are better than central (string) inverters, but for most PV rooftop systems micro-inverters have distinct advantages.

Additionally, homeowners who are subject to shading issues should definitely consider micro-inverters for their system, as they'll likely perform better compared to installing a central inverter. An analysis of the comparative advantages and disadvantages of the two systems will usually come out in favor of micro-inverters. For example, one central inverter would normally cover the requirements of an entire residential PV rooftop solar system (assuming that the central inverter has enough capacity for your entire array). Micro-inverters, on the other hand, sit on the back of every solar panel and offer several important benefits over central inverters.

For each individual PV system, the homeowner will want to determine whether the benefits of using micro-inverters outweigh the extra costs. Earlier it was stated that for those installations where shading is an issue, micro-inverters are preferred over a central inverter. This is because with even a small shading problem, a central inverter will negatively affect all of the PV modules in the array, whereas with micro-inverters only the shaded modules are affected.

Ultimately, micro-inverters make the PV system more modular and easier to expand. To subsequently increase the size of your solar electric system, you can simply add one or more panels with a micro-inverter for each new panel. Also, these new additional panels can be of different wattages and even from different manufacturers, two more significant advantages.

With the latest technological advancements, the new micro-inverters being developed may be able to harvest up to 20 percent more energy than central inverters over their lifetime. That's a lot of energy! Therefore, even though they cost more initially, it appears likely that they'll recoup the difference in a relatively short time. This also means more income from feed-in tariffs for the PV system owners who have a grid-tied or hybrid system.

Micro-inverters make a rooftop PV system or a large commercial installation more efficient. They also make it easier to monitor the power generation of each panel and enable a service technician to quickly pinpoint a failure in the system.

Micro-inverters are gaining acceptance in the solar energy market, particularly in residential applications. The market leader for many years has been Enphase, but there are a number of companies who claim to be at various stages of developing better and cheaper micro-inverters. Two relatively new companies building micro-inverters are Enecsys and SolarBridge. These two companies appear determined to be competitive; they both initially offered longer warranties than Enphase. In response, Enphase has extended their warranty period.

GreenRay, another new micro-inverter manufacturer, recently developed an interesting innovation. They have fully integrated a micro-inverter into the solar panel. Another new competitor is InfluxGreen. Their IGSM series of grid-tied PV micro-inverters (single phase 200W to 270W) appear to be cost-competitive, and the manufacturer states they can achieve a peak efficiency of up to 95 percent.

An Influx inverter—view from the front.

Although the market is rapidly expanding, it's not likely to be large enough to support all of the existing and proposed micro-inverter competitors. The market is becoming overly competitive, and some of the smaller companies will likely drop out of the race at some point. However, this need not worry the homeowner, because different makes of micro-inverters can operate together in the same array of PV panels without any compatibility problems.

An Influx inverter—view from the back.

Micro-inverters optimize output for each solar panel—not for the entire system, as a central inverter would. This enables every solar panel to perform at its maximum potential. In other words, one solar panel alone cannot drag down the performance of the entire solar array, whereas a central inverter can only optimize output according to the weakest link. For example, shading of as little as 9 percent of a solar system connected to a central inverter can lead to a system-wide decline in power output by as much as 50 percent. If one solar panel in a string had abnormally high resistance due to a manufacturing defect, the performance of every solar panel connected to that same central inverter would suffer.

One of the tricky things about solar cells is that the voltage needs to be adjusted to the right level for maximum output of power. In other words, the performance of a solar panel is dependent on the voltage load that's applied to the inverter. Maximum power point tracking, or MPPT, is a technique used to find the right voltage, or the "maximum power point." With micro-inverters, MPPT is applied to each individual panel as opposed to the solar system as a whole; performance will naturally increase.

Unlike central inverters, micro-inverters are not exposed to high power voltages or to high heat loads. Therefore, they tend to last significantly longer. Nowadays, micro-inverters typically come with a warranty of 25 years, averaging ten years longer than central inverters. The 25-year warranty for micro-inverters conveniently matches the warranty period of the solar modules, particularly if the micro-inverters are installed into the PV solar modules at the panel manufacturer's plant.

Central inverters come in limited sizes, and you might end up having to pay for an inverter bigger than what you actually need.

I recommend that you purchase or ask your solar contractor to install a monitoring system to follow the energy output of your entire PV array. But remember, with a central inverter you cannot see the output at the individual panel level. With micro-inverters, you can connect each panel to a computer monitoring system so you can easily pinpoint which panel is experiencing lower efficiency rates. This is helpful in discovering faulty panels or equipment. With micro-inverters, web-based monitoring on a panel-by-panel basis is usually available for both the homeowner and the installer. The DIY installer should invest the small extra amount required for a monitoring system. Continually analyzing the health of your PV solar rooftop system can pave the way for additional performance improvements. There are even mobile applications that enable you to monitor your PV system when you're travelling.

The main electrical panel for the house. In this case, there is a grid-connection via a back-fed breaker in the panel. While this does serve as a disconnecting means for the panel, utilities generally require a separate, accessible, dedicated AC disconnect switch such as those illustrated earlier in this chapter under the heading "DC and AC Disconnect Switches."

Solar panels are connected in a series before they're fed into a central inverter, typically with an effective nominal rating of 300-600 VDC (volts of direct current); the larger central inverters can reach up to 1,000 volts DC. Thiscurrent is potentially life threatening if safety precautions are ignored—for instance, if the DC disconnect switch is left in the "ON" position while maintenance is being done on the system. You should always refer to the safetydata sheets for the module and the inverter and follow all safety instructions.

However, micro-inverters eliminate the need for high-voltage DC wiring, whichimproves safety for solar installers and system owners—an important consideration, especially for DIY installers. Once again, we recommend readingthe safety data sheet of whatever type of inverter you plan to install.

Micro-inverters also generate significantly less heat than central inverters do, and as a result there's no need for active cooling. This enables micro-inverters tooperate without any appreciable noise, another advantage over central inverters.

Regarding the higher initial cost, it must be said that for any given PV system, micro-inverters are a more expensive option than using a central inverter. But micro-inverters are definitely worthy of consideration due to the superior long-term benefits. In recent years, central inverters had an average costof about $0.40/Wp (watts–peak), while the average cost of micro-inverters has been about 30 percent higher: $0.52/Wp. If the micro-inverters are already incorporated into the PV modules, then you'll simply compare the combined cost of the modules and micro-inverters versus the cost of the central inverter plus the cost of the PV modules.

Finally, installing solar panels with micro-inverters is simpler and less timeconsuming, which typically cuts about 10 percent off the installation costs.

Dual Micro-Inverters

A few years ago, dual micro-inverters were introduced to the market. They essentially do the same thing as regular micro-inverters, but they convert the DCof two solar panels instead of one. This lowers the initial system cost slightly, but at the price of performance, so there may be very little net benefit. A homeowner will often ask the contractor, "Are micro-inverters, or dual micro- inverters, or a central string inverter the best choice in my situation?" It depends on the site conditions, the homeowner's budget, and other factors. However inmost situations, regular micro-inverters should be given serious consideration.

However, it's only fair that we touch on the characteristics of central inverters before you make your final decision. First, central (or string) invertersare less expensive initially and have fewer moving parts when comparing the two systems overall. But in the long run, micro-inverters are usually more economical.

Still, the original single central inverter is very popular among homeowners and investors alike. The main reasons are familiarity and trust: Central inverters have simply been on the market longer, and they're believed to be efficient sincethey have a history of proven results. A typical central inverter has a maximum efficiency rate of 95 percent, and if there are no shading issues, they perform well. Because of this, they hold great promise for large industrial-and utility- sized projects, because solar systems designed for those projects usually don't confront shading challenges. For these larger installations, central inverters are significantly less expensive than micro-inverters.

Central inverters have only one point of failure. An analogy might be a ceiling with ten track lights versus a ceiling with a single lamp. Over the past year or so, there have been many claims about the reliability or unreliability ofone technology over the other, but more time is needed to determine conclusively whether central inverters or micro-inverters are more reliable. You can easily guess which special interest group is making which claim. In any case, each installation should be analyzed separately. Both inverter systems havea valid role to play, depending on the site conditions, the size of the PV system, financial considerations, and desired system flexibility.

Though both central inverters and micro-inverters have a place in the market,micro-inverters are gaining ground and have become the inverter of choice for many residential PV solar installation companies throughout the US and overseas. We can evaluate the benefits of micro-inverters versus central invertersby looking at two numbers:

- Lifetime costs ($)
- Lifetime energy production (kWh)

These two figures, calculated for both types of inverters, are essential numbers. Divide costs by energy production and you can determine how much money you'll pay for every kWh your solar system will produce. Every situation is different. There are several variables to take into account in order to find these two numbers.

Enecsys, one of the leading micro-inverter manufacturers, sums it up like this: "A total cost-of-ownership analysis of a PV solar system can only be carried out after detailed examination of capital and maintenance costs, and an understanding of how much energy will be harvested over the life of the system."

However, if your solar contractor recommends using a standard central inverter for your installation, and some good reasons are provided, this is fine. The central inverter is a readily available and well-proven system.

Your central or string inverter unit is not much bigger than a typical disconnect switch, and they can be installed side-by-side for convenience and an aesthetically pleasing appearance. Both the disconnect switch and the inverter can be installed close to your main breaker panel or breaker box. Between the local grid line transformer and the main breaker panel, your electrician will install a "smart meter," which will measure in real time the electric power provided by the PV solar system to your residence. If you have selected an in-grid system, the smart meter will also be connected to the external power grid, probably near the existing conventional electric meter.

Below are three schematic diagrams of the different available metering systems, "FiT," "PPA," and "Net Metering":

The smart meter will measure and record the power in kilowatt hours consumed by your home, as well as the excess amount of power fed back into the external grid of your utility company. Thus, the term "net metering" is used to describe the net amount of power consumed. As mentioned before, most utility companies will pay you in the form of credit for any excess daily power generated from your PV system. For domestic consumers, there's usually an upper limit to these credits that is equal to the charges for power consumed. So, theoretically, you could have some months with a net metering month-end bill of zero. To qualify for net metering, your PV system must be less than a specified maximum generating capacity, which is regulated by your electric utility. In the US, the most common maximum size cap is 10 kW for residential systems.

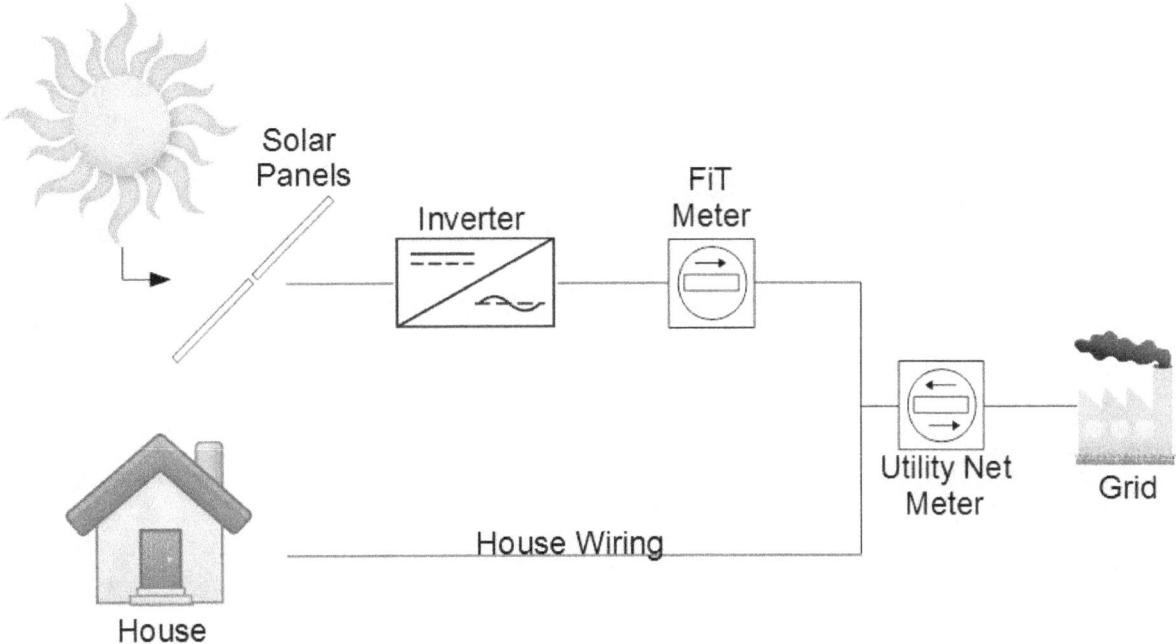

A schematic of a FiT meter connection..

The location of your existing breaker panel box may be where you install the PV solar DC disconnect switch, inverter, and smart meter. The breaker panel is usually installed on a wall in a garage, or in a hallway near the back door, or on a wall outside of the house.

Power moves from the DC/AC inverter to your home breaker panel box and is distributed to the rest of the houschold. A power meter with net metering capability is a little different from the standard meter you have now. It's capable of measuring power going into the grid or being pulled from the grid at the end of the line, and it will measure the amount of electricity that is either consumed or being sold back to the utility company.

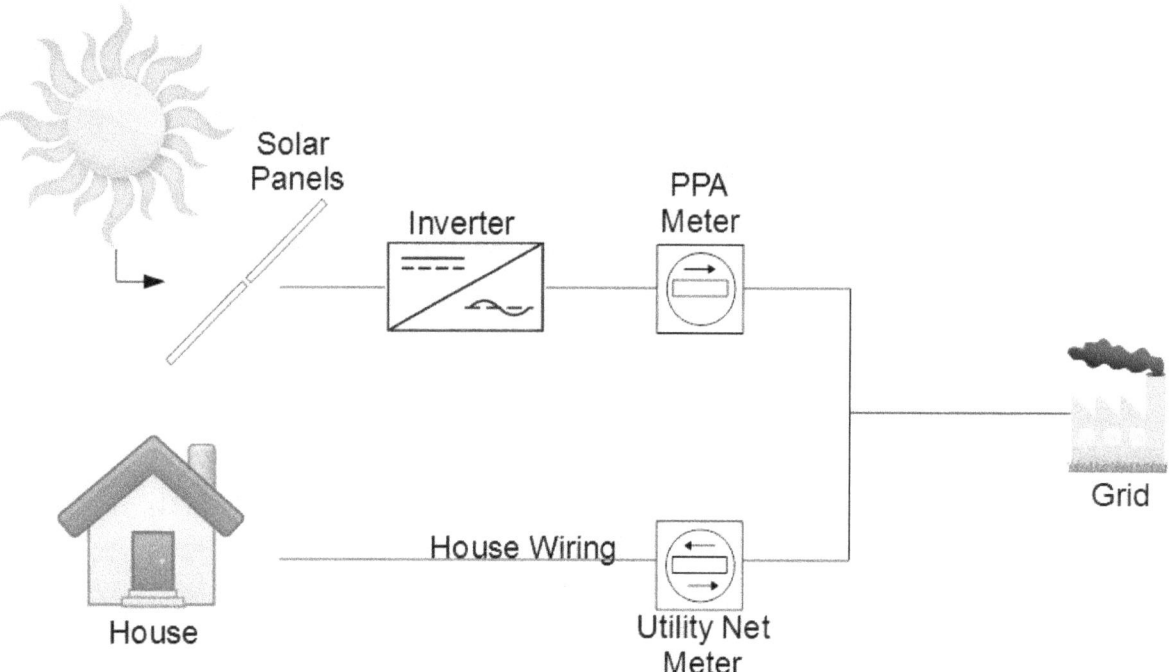

A schematic of a PPA meter connection.

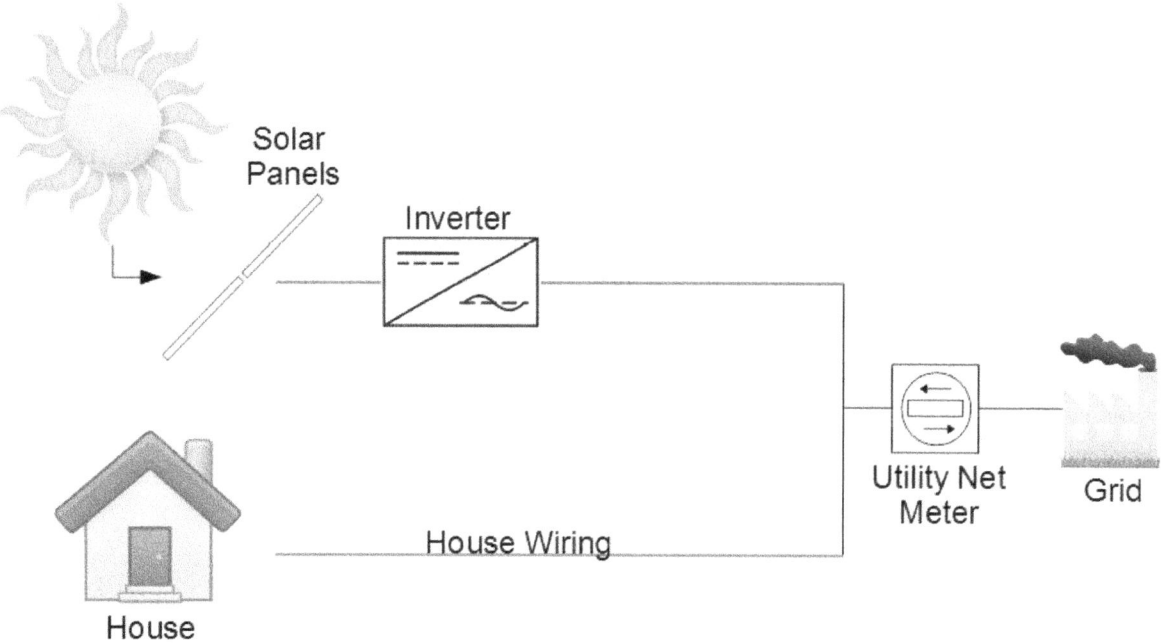

A schematic of a net metering connection.

With net metering, safety is an issue as well. The utility has to make sure that if there's a power outage in your immediate area or neighborhood, your PV system won't continue to feed electricity into power lines that a lineman might think are dead. This is a dangerous situation

64

called "islanding," but it can be avoided with an anti-islanding inverter. Most inverters incorporate the anti- islanding protection feature.

The thought of living at the whim of the weatherman probably doesn't thrill most people, but three main options can ensure you still have power even if the sun isn't cooperating. If you want to live completely off the grid but don't trust your PV panels to supply all the electricity you'll need in a pinch, you can use a backup portable diesel generator when sunlight is low or if you wish to have power after dark or during a blackout in your area.

The second stand-alone system involves energy storage in the form of batteries. Several batteries connected together form what is commonly called a "battery bank." Unfortunately, batteries can add a lot of cost and maintenance to a PV solar rooftop system. But installing a battery bank is a necessity if you want to be completely independent of the electric utility company grid.

However, if you find that the cost of a battery bank and backup generator large enough for your requirements is excessive, then the alternative is to connect your PV solar system to the utility grid, assuming it's available, buying power when you need it, and selling it back to the utility when your PV system produces more power than you use. A more detailed description and several images of battery banks are provided in Chapter 2.

If you decide to use batteries instead, keep in mind that they'll have to be maintained, and they must be replaced after a certain number of years. Most solar panels tend to last at least 30 years—and improved longevity continues as a main research goal—but batteries just don't have that kind of useful life.

Depending on various factors, solar batteries might last up to 10 or 11 years before they must be replaced. Also, be aware that PV battery bank systems can be potentially dangerous because of the energy they store and the acidic electrolytes they contain, so you'll need a well-ventilated space and a nonmetallic enclosure or rack where they can operate safely.

Although several different kinds of batteries are commonly used for PV systems, make sure they're deep-cycle batteries. Unlike your car battery, which is a shallow-cycle battery, deep-cycle batteries can discharge more of their stored energy while still maintaining long life. Car batteries discharge a large current for a very short time to start the car engine, and then the alternator immediately recharges them as you drive. PV solar batteries generally have to discharge a smaller current for a much longer period of time, such as at night or during a power outage, while being charged during the sunny parts of the day. The most commonly used deep-cycle batteries are lead-acid batteries (both sealed and vented) and nickel-cadmium batteries, both of which have various pros and cons. Refer to Chapter 2 for more information, and, as recommended above for inverters, be sure to read the safety data sheet, or SDS, that comes with every important piece of electrical equipment. Any distributor or contractor can provide you with the SDS for any piece of equipment they sell or install.

Other Components Needed for Your PV Solar Rooftop System

CHARGE CONTROLLER

The use of batteries requires the installation of another component called a "charge controller." Batteries last a lot longer if they aren't overcharged or drained too much. That is one of the functions of a charge controller. Once the batteries are fully charged, the charge controller doesn't let current from the PV modules continue to flow into them. Similarly, once the batteries have been drained to a certain predetermined level of voltage, the charge controller won't allow more current to be drained from the batteries until they have been recharged. The use of a charge controller is essential for long battery life.

The other challenge, besides your energy storage system, is that the electric current generated by your solar panels—or extracted from your batteries, if you choose to use them—is not in the form that is supplied by the utility, namely AC, which is also the form of current required by the electrical appliances in your home. A solar system generates DC, so you need an inverter to convert it into alternating current. As discussed earlier, apart from switching DC to AC, most inverters are also designed to protect against islanding if your system is hooked up to the power grid.

Most large inverters will allow you to automatically control how your system works. Some PV modules, called AC modules, actually have a small inverter already built into each module, eliminating the need for a central inverter and simplifying wiring issues; refer to the comparison of inverters and micro- inverters above.

In addition to your PV array of panels and mounting support hardware, the wiring, conduits, junction boxes, grounding equipment, over-current protection, DC and AC disconnect switches, net metering equipment, and other accessories identified above will round out the essential components of a PV solar system. Of course, both DIY installers and solar contractors must follow regulations governed by electrical codes; there is a section in the National Electrical Code just for PV solar systems. There may also be state regulations or municipal regulations that apply to the installation of your PV solar rooftop system.

I recommend that you contact a licensed electrician who has experience with PV solar systems to connect the electric control equipment and measuring devices during your installation. In Chapter 9, I provide links to websites that supply contact information for experienced solar contractors and PV electricians, information that will help you contact experienced solar contractors in your hometown or nearby. Generally speaking, solar contractors are willing to let DIY enthusiasts handle much of the PV solar rooftop installation process, including mounting the support system, the PV modules, and the conduits.

Once installed, a PV solar rooftop system requires very little maintenance, especially if no batteries are used, and will provide electricity cleanly and quietlyfor 25 to 30 years.

Chapter 4 - How to Determine the Optimal Size of Your System

You need to need to know a few basic things before you can proceed to design the layout for your solar energy system. First of all, you need to determine how much energy you want your proposed rooftop PV solar system to produce. If you're going to install a PV solar system on the rooftop of your existing home, and you aren't contemplating building any additional rooms, then it's an easy task to determine your electric consumption by analyzing your monthly utility bills. You can simply record and add up the kilowatt hours (kWh) during the previous 12 months and calculate an average. For most single-family dwellings, this will be in the range of 7,000 to 9,000 kWh per month. This number can vary a lot, depending on the size of the house, number and ages of the children, consumption habits of all family members, and weather conditions.

If you're building a new home, you won't have any electric bills to indicate a monthly consumption level. In this case, you'll need to do some calculations to determine your future electric consumption by adding up the estimated consumption of all the appliances, lights, etc., and estimating the number of hours per 24-hour day each item will be in use on average. Approximations or rough estimates will suffice.

Step 1: Determine Your Power Consumption Requirements

Make a list of the appliances and other electrical devices you are planning to power from your PV system. Using the table here as a guide, find out how much power each electrical item in the house consumes when it's operating or turned on. Most appliances and other electric or electronic devices have a label on the back that lists the wattage. Specification sheets, local appliance dealers, and product manufacturers are other sources of information. For your convenience, the chart here lists typical power consumption levels or wattage ratings of common household appliances and devices. With the wattage ratings for each electric appliance and other power consuming devices, you will be ready to fill out a "load-sizing worksheet."

Typical Power Consumption of Household Appliances and Other Domestic Electric Devices	
APPLIANCE OR ELECTRIC DEVICE	**WATTS**
Blender	300
Blow Dryer	1000
CB Radio	5
CD Player	35
Ceiling Fan	10–50
Clock Radio	1
Clothes Dryer (electric)	400
Clothes Dryer (gas heated)	300–400
Coffee Pot	200
Coffeemaker	800
Compact Fluorescent Incandescent (40 watt equivalent)	11
Compact Fluorescent Incandescent (60 watt equivalent)	16

Compact Fluorescent Incandescent (75 watt equivalent)	20
Compact Fluorescent Incandescent (100 watt equivalent)	30
Computer Laptop	20–50
Computer PC	80–150
Computer Printer	100
Electric Blanket	200
Electric Clock	3
Freezer, 14cf (15 hours)	440
Freezer, 14cf (14 hours)	350
Freezer, 19cf (10 hours)	112
Hot Plate	1200
Iron	1000
Light Bulb (100W Incandescent)	100
Light Bulb (25W Compact Fluorescent)	28
Light Bulb (50W DC Incandescent)	50
Light Bulb (40W DC Halogen)	40

Light Bulb (20W Compact Fluorescent)	22
Microwave	600–1500
Refrigerator/Freezer, 20 cf 1.8 kwh per day (15 hours)	540
Refrigerator/Freezer, 16 cf 1.6 kwh per day (13 hours)	475
Refrigerator, 16cf DC (7 hours)	112
Refrigerator, 12cf DC (7 hours)	70
Satellite Dish	30
Sewing Machine	100
Shaver	15
Stereo System	10–30
Table Fan	10–25
Toaster	800–1500
TV (25")	150
TV (19")	70
Vacuum Cleaner (upright)	200–700
Vacuum Cleaner (hand)	100
VCR	40

Washing Machine (automatic)	500
Washing Machine (manual)	300
Water Pump	250–500

Step 2: Load-Sizing Procedure

First, list all of the electrical appliances to be powered by your PV system. Then separate AC and DC devices, if applicable, and enter them in the table. Then insert a column for the number of items (more columns, if you have two or more of the same appliance). Next, you can record the operating wattage of each item. Then you need to specify the approximate number of hours per day each item will be used. Next, you multiply the three numerical columns to determine average daily watt–hour usage per line item. Then you need to insert another column and enter the number of days per week you'll be using each item to determine the total watt–hours per week each appliance will require. Most appliances have a label on the back that lists the wattage. Local appliance dealersand product manufacturers are other sources of information.

Finally, add all the numbers in column E and insert that number into the bottom right-hand cell. Excel can add these numbers for you automatically: Highlight all the cells to be added horizontally (including the last cell on the bottom row), then simply click the cursor on the "sum" function key in the topmenu of the Excel spreadsheet. The sum total will immediately appear in thebottom right hand cell.

If you have any DC appliances or devices, you can do a separate small spreadsheet. Or you can include them in the same table. What you want is the "total watts per week" (column E), regardless of whether the items are AC or DC. In any case, most homeowners will not have any DC appliances.

You can easily adapt the columns and rows to your specific situation.

Step 3: Size Your Battery Storage System

Many homeowners will elect to go with a grid-tied PV system and net metering, relying on grid power to supply all electricity that can't be supplied by the solar panels, including at night and during the day when it's raining or heavy cloud conditions prevail. These homeowners are satisfied to use their solar PV system as much as sunlight hours will allow and to rely on the utility grid to make upthe difference.

These homeowners will normally decide not to install a battery storage system because of the substantial increase it would represent to the total cost of the system. In addition, batteries require more maintenance than PV panels and batteries have a shorter life. After eight or ten years, they must be replaced with new batteries at substantial cost.

Nonetheless, there are a number of homeowners who want a battery storage system, either because their home does not have access to the utility grid, or because their goal is to be totally independent of grid power.

If, for whatever reason, you're definitely interested in a battery storage system, you should feel comfortable choosing an appropriate deep-cycle battery to use as a backup energy storage system. After reading this chapter, you'll be ready to fill out a battery sizing worksheet.

Load-Sizing Worksheet (Part 1)					
	A	B	C	D = A × B × C	E = D × 7
APPLIANCE OR DEVICE	WATTS	QUANTITY	AVG HRS/DAY	= WH/DAY	= WH/WK
1)					
2)					
3)					
4)					
5)					
Continue the list until complete					
TOTAL	n/a	n/a	n/a	n/a	

(The easiest way to do this is using a simple Excel worksheet.)

The first decision you'll need to make is how much energy storage (average daily power consumption multiplied by the reserve power time in days) you would like your battery bank to provide. This means how many consecutive days you want your PV system to power your home without using the utility grid power, or how many consecutive cloudy or rainy days do you want to operate on your battery system. Often, this is expressed as "days of autonomy," because it's based on the number of days you expect your system to provide power without receiving an input charge from the solar array. In addition to days of autonomy, you should also consider your usage or consumption patterns and any significant differences in your consumption needs for different seasons.

If you're installing a system for a weekend home, you might want to consider a larger battery bank, because your system will have all week to charge and store energy. Alternatively, if you're adding a PV array as a supplement to a generator-based system, your battery bank can be slightly smaller, since the generator can be operated if needed during the recharging cycles.

Batteries are sensitive to temperature extremes, and you cannot take as much energy out of a cold battery as a warm one. However, you can use the chart provided to factor in temperature adjustments. Although you can get more than rated capacity from a hot battery, operation at very hot temperatures will shorten battery life slightly. Try to keep your batteries near room temperature. Charge controllers can be purchased with a temperature compensation option to optimize the charging cycle at various temperatures and lengthen your battery life.

Under "Grid-Tied Solar Systems", I stated that the depth of discharge, "DOD," is defined as the minimum charge level at which point your PV system stops draining the battery and starts the recharge cycle again. The DOD can also be referred to as the percentage of the rated battery capacity (ampere-hours) that can safely be extracted from the battery before a recharge cycle is needed. The capability of a battery to withstand discharge depends on its construction. The terms "shallow-cycle" and "deep-cycle" are commonly used to describe the two main types of batteries. Shallow-cycle batteries are lighter, less expensive, and have a short lifetime. For this reason, deep-cycle batteries should always be used for stand-alone PV systems. These units have thicker plates and most will withstand daily discharges of up to 80 percent of their rated capacity, although most battery manufacturers recommend a maximum discharge of 50 percent. Most deep-cycle batteries are referred to as "flooded electrolyte." This means that the plates are covered with the electrolyte solution, so the level of fluid must be monitored and distilled water must periodically be added to keep the plates fully covered. If this doesn't appeal to you, go with sealed lead-acid batteries, which don't require liquid refills. For special applications there are other types of deep-cycle batteries, such as nickel cadmium, as discussed in Chapter 2.

The depth of discharge value used for sizing should be the worst case scenario, or the maximum expected discharge that the battery will experience— for example 50 percent, 55 percent, or 60 percent. The system control unit, or charge controller, should be set to prevent discharge beyond whichever specified level you choose.

The ampere-hour (Ah) capacity of a battery is usually specified as well, together with a standard-hour reference such as 10 or 20 hours. For example, suppose the battery is rated at 100 ampere-hours and a 20-hour discharge time is specified. When the battery is fully charged, it can deliver a discharge current of 5 amperes for 20 hours (5A × 20 hrs = 100 ampere-hours). If you're ever confused or have doubts about your understanding of the relationship between the capacity of a battery and the load current, information is always provided in the literature that comes with every battery and on the website of the battery manufacturer.

Because it is dependent on factors such as charge rate, discharge rate, depth of discharge, number of cycles, frequency of use, and operating-temperature extremes, the precise lifetime of any battery is difficult to predict. It would be quite unusual for a lead-acid battery to last longer than 11 or 12 years in a PV system, but some may last longer, depending on how often they're used and how they are maintained.

Batteries require periodic maintenance. Even a sealed battery should be checked to make sure connections are tight and there is no indication of overcharging. For flooded batteries, the electrolyte level should be maintained above the plates and the voltage of the battery kept at the proper charge level. The specific gravity of each cell should also be checked to ensure consistent values between cells. Wide variations between the cell readings would probably indicate problems. The specific gravity of the cells should be checked with a hydrometer and recorded, particularly before the onset of winter in temperate climates.

In very cold environments, the electrolyte in lead-acid batteries may freeze. The freezing temperature is a function of the battery's state of charge. When a battery is completely discharged, the electrolyte essentially becomes water, and the battery may be subject to freezing. If you live in a climate with very cold winter months, try to keep your battery bank in a room that doesn't get too cold.

Now you're ready to calculate how many batteries you'll need for your system, based on your own particular criteria. Just fill in the blanks. It'll be easy with the information, tables, and formulas discussed above.

If a battery bank of this size is too costly, or if it requires more space than you have available, you might want to consider reducing the size of your battery bank by reducing the daily amp-hour requirement (line 1, above). This is logically more feasible for a hybrid system. And since battery storage systemsare modular, it's relatively simple to increase the size of your battery bank later, if you want.

How to Use Your Local Insolation (or Solar Irradiance) Index and Power Consumption Requirements to Determine the Optimal Size of a PV Array for Your Home

You'll need to determine how many solar panels can fit on your roof, how muchsolar radiation you receive at your location, your average monthly power consumption requirements, and how big a system you can afford or wish to buy.An array can be planned depending on any or all of these considerations, but taking all of these factors into account will be the wisest approach.

Worksheet for Calculating the Number of Batteries You Require		
ITEM	FORMULA OR DESCRIPTION	ANSWER/UNIT
1	Enter here the daily power consumption or amp-hour requirement. (Obtain this number from the Load Sizing Worksheet above, line 5).	_____Ah/day
2	Enter here the number of days you want your system to provide power autonomously, without your PV panels receiving any charge (that is, the number of continuous days of rain or overcast weather).	_____days
3	Multiply the amp-hour requirement (line 1) by the number of days of autonomy desired (line 2). This is the amount of amp-hours your battery system will need to store.	_____Ah
4	Enter the depth of discharge for the battery you have chosen. This is asafety measure used to avoid excess discharge of the battery bank. (Example: If the DOD you chose is 50 percent, use 0.5 here. Your DOD should not be more than 0.8).	_____(e.g. 0.5 to 0.8)
5	Divide the amp-hours needed (line 3) by the DOD (line 4). This is the number of Ah required if the ambient temperature is above 80°F / 26.7°C.	_____Ah
6	Enter the "temperature factor multiplier." Use the estimated lowestambient winter temperature of the room or space where your batterybank is to be located. Consult the list of temperature factor multipliers below and use the entry that corresponds with your estimated lowest ambient winter temperature.	_____ [temperature factor multiplier]

77

7	Multiply line 5 by line 6. This will be the total energy storage capacity required by your battery bank.	_____ = _____ Ah (Total)
8	Enter the amp-hour rating of your individual batteries. You can find this on the spec sheet that comes in each battery box. Use the 20-hour rating provided on the spec sheet.	_____ batteries
9	Divide the total energy storage capacity (line 7) by the amp-hour rating for 20 hours of usage. Round up to the next whole number. This will be the number of batteries needed to be wired in parallel.	_____ batteries
10	Divide the nominal system voltage (12V, 24V, or 48V) by the individual battery voltage. This will be the number of batteries to be wired in the series. Round up to the nearest whole number.	_____ batteries
11	Multiply line 9 by line 10. This will give you the total number of batteries you need for your chosen battery bank.	_____ batteries

Use the Temperature Factor Multiplier for line 6. Pick the lowest winter temperature where your batteries will be installed. A simple estimate will suffice.

Temperature Factor Multiplier	
LOWEST AMBIENT WINTER TEMPERATURE	**MULTIPLIER**
80°F 26.7°C	1.00
70°F 21.2°C	1.04
60°F 15.6°C	1.11
50°F 10.0°C	1.19
40°F 4.4°C	1.30
30°F −1.1°C	1.40

20°F −6.7°C	1.59

Knowing how much power you currently consume on average is the best starting place if you already own your own home; this information is readily available from your monthly electric bills. With this information, you can estimate the array size that will produce the same amount of energy on a monthly basis, how much energy can be produced in the space available, or how much energy you can produce on the budget you have allowed for the project.

Once you know what your target monthly output is in kilowatt hours (kWh), simple math will let you calculate how big an array you need. Then you can focus on choosing the type of solar panels you want, where to install them (the layout or configuration), and how many panels you'll need.

When looking at your utility bills, don't forget that your electricity consumption changes with the seasons, especially in temperate climates and countries that are subject to cold winters and hot summers. Most homes in the northern US and in most of continental Europe use 700–900 kWh each month, but this consumption level can differ substantially according to the country, the season, the type and size of home, family size, energy efficiency of appliances and lighting fixtures, and the family's energy consumption habits.

You can divide your expected monthly consumption level in kWh by the peak sun hours you receive each day to determine the size of your system. Peak sun hours are the number of hours each day when the insolation (or solar irradiance) equals 1000 watts/square meter. Information on peak sun hours in nearly every town and city in the world is readily available on the Internet. In the United States, for example, you can find out how much sunlight you receive per day on the National Renewable Energy Laboratory website under the "US Solar Resource Map" for photovoltaics, at www.nrel.gov/gis/solar.html.

For a clearer picture, let's take the example of 900 kWh/month, which is 30 kWh/day. If you use 30 kW daily, and you receive an average of four hours of sunlight per day, and if the PV panels you selected have a rated capacity of 250 Wp (watts peak), you'll need to make an adjustment because the panels' capacity is measured in direct current whereas the electricity we use in our homes and offices is always alternating current, and the DC produced by the PV panels is converted into AC power by an inverter. This process by which AC is converted to DC involves the loss of a certain amount of power. This loss is calculated by multiplying the Wp in DC by a number we call the "derate factor" to get the equivalent power output in AC. The derate factor may vary a little, depending on the system and the inverter used, but a safe average derate factor we can use for calculating our system is 0.82. This means that your PV panel rated at 250 Wp DC will actually deliver 250 × 0.82 = 205 watts AC. So in our example, this home is consuming 30 kW daily = 30,000 Wh/day (AC).

A typical installation of poly modules that I estimate will have an installed capacity of 900 kWh/month, depending on the location of where it is installed.

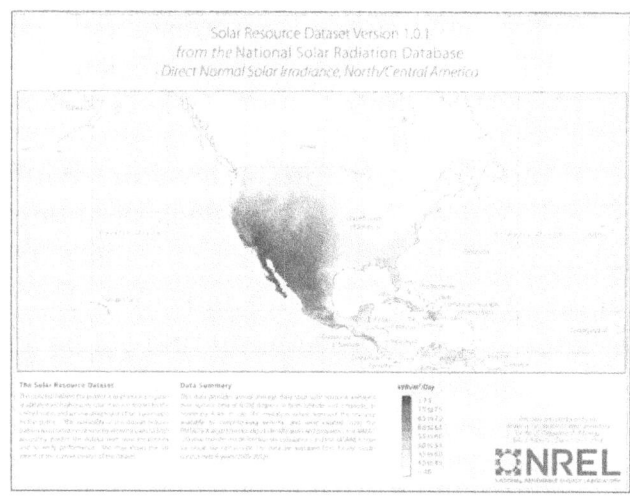

This map provides the solar irradiance levels for all areas in the United States, Canada, and Central America.

Therefore, divide 30,000 Wh/d by the average hours of sunlight available at your location, otherwise known as the local insolation index. For simplicity, many people will use 4 hours per day. But we want to be more precise, so you'll need to use the exact insolation index number for your location, as given in the table below (for Europe and other areas, we refer to other tables on the Internet). Let us use the example of Los Angeles, California, which has an insolation idex of 5.62 (yearly average). So we divide 30,000 Wh/d by 5.62 h/d, which gives us 5,338 W. Then this number is divided by the adjusted PV panel power delivery rating of 205 W/panel (AC). Thus, the calculation is 5,338 ÷ 205 = 26.0 panels. This means that the array will require 26 panels to produce power equivalent to the home's daily average consumption of 30 kW/d, using the PV modules specified above and rated at 250 Wp (DC).

Step 4: Determine the Average Sun Hours Available Per Day

Several factors, including seasonal usage, typical local weather conditions, fixedmountings versus trackers, and location and angle of the PV array, influence how much solar power your modules will be exposed to.

The following table provides the number of hours of full sunlight available togenerate electricity. Your solar array power generation capacity is dependent on the angle of the sun's rays as they hit the modules. Peak power occurs when the sun's rays are at right angles, or perpendicular, to the modules. As the angle deviates from perpendicular, more and more of the energy is reflected rather than absorbed by the modules. Depending on your application and your budget, sun-tracking mounts can be used to enhance your power output by automatically adjusting the position of the panels to track or follow the sun while maintaining a 90-degree angle.

The solar insolation table below (also known as the solar irradiance) tells us the amount of sunlight available during spring, summer, autumn, and winter, measured in peak sun hours or simply "sun hours." It's more difficult to produce energy during the winter because of shorter days, increased cloudiness, and thesun's lower position in the sky. This table lists the sun-hour ratings for over 75 cities in North America for three categories: summer, winter, and yearly average.

If you use your system primarily in the summer, use the summer value. If you're using your system year-round, especially for a critical application, use thewinter value. And finally, if you're using the system most of the year (spring, summer, and fall) or the application is not critical, use the average value.

Wherever you live in North America or overseas, you should be able to determine a reasonable estimate of the sun's availability (peak hours) in your area by using the map above and the table below. The map provides color codingmatched with index numbers that indicate the general solar insolation index number for anywhere on the globe.

Solar Insolation or Solar Irradience Levels (Peak Sun Hours) for Cities in the USA and Canada			
USA BY STATE, CITY	SUMMER AVG.	WINTER AVG.	YEAR AVG.
AL, Montgomery	4.69	3.37	4.23
AK, Bethel	6.29	2.37	3.81
AK, Fairbanks	5.87	2.12	3.99

AK, Mantanuska–Susitna Borough	5.24	1.74	3.55
AZ, Page	7.3	5.65	6.36
AZ, Phoenix	7.13	5.78	6.58
AZ, Tucson	7.42	6.01	6.57
AR, Little Rock	5.29	3.88	4.69
CA, Davis	6.09	3.31	5.1
CA, Fresno	6.19	3.42	5.38
CA, Inyokern	8.7	6.97	7.66
CA, La Jolla	5.24	4.29	4.77
CA, Los Angeles	6.14	5.03	5.62
CA, Riverside	6.35	5.35	5.87
CA, Santa Maria	6.52	5.42	5.94
CA, Soda Springs	6.47	4.4	5.6
CO, Boulder	5.72	4.44	4.87
CO, Granby	7.47	5.15	5.69
CO, Grand Junction	6.34	5.23	5.86
CO, Grand Lake	5.86	3.56	5.08
DC, Washington	4.69	3.37	4.23

FL, Apalachicola	5.98	4.92	5.49
FL, Belle Island	5.31	4.58	4.99
FL, Gainesville	5.81	4.71	5.27
FL, Miami	6.26	5.05	5.62
FL, Tampa	6.16	5.26	5.67
GA, Atlanta	5.16	4.09	4.74
GA, Griffin	5.41	4.26	4.99
HI, Honolulu	6.71	5.59	6.02
IA, Ames	4.8	3.73	4.4
ID, Twin Falls	5.42	3.41	4.7
ID, Boise	5.83	3.33	4.92
IL, Chicago	4.08	1.47	3.14
IN, Indianapolis	5.02	2.55	4.21
KS, Dodge City	4.14	5.28	5.79
KS, Manhattan	5.08	3.62	4.57
KY, Lexington	5.97	3.6	4.94
LA, Lake Charles	5.73	4.29	4.93
LA, New Orleans	5.71	3.63	4.92

LA, Shreveport	4.99	3.87	4.63
ME, Blue Hill	4.38	3.33	4.05
MA, Boston	4.27	2.99	3.84
MA, E. Wareham	4.48	3.06	3.99
MA, Lynn	4.6	2.33	3.79
MA, Natick	4.62	3.09	4.1
MD, Silver Hill	4.71	3.84	4.47
ME, Caribou	5.62	2.57	4.19
ME, Portland	5.2	3.56	4.51
MI, E. Lansing	4.71	2.7	4.0
MI, Sault *Ste.* Marie	4.83	2.33	4.2
MN, St. Cloud	5.43	3.53	4.53
MO, Columbia	5.5	3.97	4.73
MO, St. Louis	4.87	3.24	3.78
MS, Meridian	4.86	3.64	4.44
MT, Glasgow	5.97	4.09	5.15
MT, Great Falls	5.7	3.66	4.93
NC, Cape Hatteras	5.81	4.69	5.31

NC, Greensboro	5.05	4.0	4.71
ND, Bismarck	5.48	3.97	5.01
NE, Lincoln	5.4	4.38	4.79
NE, North Omaha	5.28	4.26	4.9
NJ, Seabrook	4.76	3.2	4.21
NM, Albuquerque	7.16	6.21	6.77
NV, Ely	6.48	5.49	5.98
NV, Las Vegas	7.13	5.83	6.41
NY, Bridgehampton	3.93	1.62	3.16
NY, Ithaca	4.57	2.29	3.79
NY, New York	4.97	3.03	4.08
NY, Rochester	4.22	1.58	3.31
NY, Schenectady	3.92	2.53	3.55

| OH, Cleveland | 4.79 | 2.69 | 3.94 |
| OH, Columbus | 5.26 | 2.66 | 4.15 |

OK, Oklahoma City	6.26	4.98	5.59
OK, Stillwater	5.52	4.22	4.99
OR, Astoria	4.76	1.99	3.72
OR, Corvallis	5.71	1.9	4.03
OR, Medford	5.84	2.02	4.51
PA, Pittsburgh	4.19	1.45	3.28
PA, State College	4.44	2.78	3.91
RI, Newport	4.69	3.58	4.23
SC, Charleston	5.72	4.23	5.06
SD, Rapid City	5.91	4.56	5.23
TN, Nashville	5.2	3.14	4.45
TN, Oak Ridge	5.06	3.22	4.37
TX, Brownsville	5.49	4.42	4.92
TX, El Paso	7.42	5.87	6.72
TX, Fort Worth	6.0	4.8	5.83
TX, Midland	6.33	5.23	5.83
TX, San Antonio	5.88	4.65	5.3

UT, Flaming Gorge	6.63	5.48	5.83
UT, Salt Lake City	6.09	3.78	5.26
VA, Richmond	4.5	3.37	4.13
WA, Prosser	6.21	3.06	5.03
WA, Pullman	6.07	2.9	4.73
WA, Richland	6.13	2.01	4.43
WA, Seattle	4.83	1.6	3.57
WA, Spokane	5.53	1.16	4.48
WV, Charleston	4.12	2.47	3.65
WI, Madison	4.85	3.28	4.29
WY, Lander	6.81	5.5	6.06
CANADA BY PROVINCE, CITY			
Alberta, Edmonton	4.95	2.13	3.75
Alberta, Suffield	5.19	2.75	4.1
British Columbia, Kamloops	4.48	1.46	3.29
British Columbia, Prince George	4.13	1.33	3.14
British Columbia, Vancouver	4.23	1.33	3.14

Manitoba, The Pas	5.02	2.02	3.56
Manitoba, Winnipeg	5.23	2.77	4.02
New Brunswick, Fredericton	4.23	2.54	3.56
Newfoundland, Goose Bay	4.65	2.02	3.33
Newfoundland, St. Johns	3.89	1.83	3.15
Northwest Territory, Fort Smith	5.16	0.88	3.29
Northwest Territory, Norman Wells	5.04	0.06	2.89
Nova Scotia, Halifax	4.02	2.16	3.38
Ontario, Ottawa	4.63	2.35	3.7
Ontario, Toronto	3.98	2.13	3.44
Prince Edward Isl., Charlottetown	4.31	2.29	3.56
Quebec, Montreal	4.21	2.29	3.5
Quebec, Sept-Isles	4.29	2.33	3.5
Saskatchewan, Swift Current	5.25	2.77	4.23

| Yukon, Whitehorse | 4.81 | 0.69 | 3.1 |

Step 5: Size the Optimal PV System for Your Home

With the information calculated above, we now have enough data to calculate the optimal size for your personally specified solar PV array, as follows: Worksheet to Calculate the Size (Total Capacity) of Your PV Solar System (Number of PV Panels and Total System Power in Watts)

1. Enter your daily amp-hour requirement (from line 5 of the "load sizing worksheet").	_____ Ah/day
2. Enter the sun-hours per day (solar irradiation index number) for your area. Refer to chart in Step 4.	_____ h/day
3. Divide line 1 by line 2. This is the total amperage required from your solar array.	_____ amps
4. Enter the peak amperage (DC) rating of the solar module you have selected.	_____ amps
5. Divide line 3 by line 4. This is the number of solar modules needed in parallel.	_____ panels (in parallel)
6. Select the required modules in series from the following chart. | BATTERY BANK VOLTAGE	NUMBER OF MODULES IN SERIES
12V	1
24V	2
48V	4		_____ panels (in series)
7. Multiply line 5 by line 6 to find the total number of modules needed in your array.	Total: _____ panels
8. Enter the adjusted nominal power rating (in watts) of the module you have chosen. This is the Wp rating multiplied by the "derate factor."	_____ watts
9. Multiply line 7 by line 8. This is the nominal power output of your personally specified PV solar rooftop system.	_____ watts

How Much Money Can We Save with PV Solar?

Instead of evaluating a solar system by its capacity in kilowatts (kW), you might want to estimate how many kilowatt-hours (kWh) any solar system can be expected to generate over its estimated lifetime. If we divide this number into the overall cost of the system and then divide the result by the number of years of useful life expected from the system (use the warranty period of 25 years), we obtain the average annual cost of the electricity produced by your proposed PV system in $/kWh/year. In this way, we can also compare equivalent annual cost figures for different PV solar systems. You're now left with how much you are paying for every kWh of electricity the solar panels produce ($/kWh). Divide again by 100 to get cents/kWh.

Electricity prices in the United States typically range from 8 to 17 cents/kWh, depending on the state or region. Therefore, you can easily calculate the annual savings produced by your PV system by taking the standard utility rate for electricity in your area (in cents/kWh) and subtracting the number you calculated above for the average cost of the electricity (also in cents/kWh) generated by your proposed PV solar system. Then, take the difference of these two figures and multiply that number by the sum of your actual 12 months of consumption (in kWh). You can extract this number from the previous 12 months of your utility bills. This will give you the average annual savings provided by your PV solar system.

If you're just comparing two different solar panels, and the length of their warranties is the same, doing a similar analysis for the first year (as opposed to their lifetime) will be sufficient. When making this comparison, it's helpful to know that nowadays, solar panels typically have a warranty period of 25 years. The performance of solar panels degrades very slowly over time, usually less than 1 percent per year. For this reason, solar panel manufacturers usually guarantee that the power output of their solar panels will stay above 75 to 80 percent until the warranty expires.

Solar System 2 is superior from a financial standpoint ($38,000 vs. $42,000), and most homeowners would choose it for this reason. System 1 is a better option for homeowners who lack roof space (375 ft^2 vs. 485 ft^2). The slightly more expensive solar system will save space at a moderately higher system cost

per kWh. Generally speaking, the less efficient and cheaper solar panels tend to produce slightly more electricity in a year for the same amount of money compared to more expensive, high-efficiency solar panels.

Case Study: Polycrystalline Modules		
DESCRIPTION	**SOLAR SYSTEM 1**	**SOLAR SYSTEM 2**
Manufacturer	SunPower	Canadian Solar
Model	SPR-210-WHT-U	CS6P-250P
Efficiency	16.9 percent	16.1 percent
Number of Solar Panels	28	28
Output (Year 1)	8,930 kWh	9,066 kWh
Total System Cost	$42,000	$38,000
Area	375 ft^2 (34.8 m^2)	485 ft^2 (45.1 m^2)
Cost per kWh	$4.70/kWh	$4.19/kWh

Note: The total system costs used above will vary considerably from region to region and may change substantially over time. Thus, the reader should not use these figures to compare with any local quotations.

Chapter 5 - How to Calculate the Annual Output of a System and the Optimal Number of Panels

The conventional formula used to estimate the electricity generated in annual output of a photovoltaic system is: Energy (kWh) = A × r × H × PR

Where:

A = Total solar panel Area (m²) r = Solar panel yield (percent)

H = Annual average solar radiation on tilted panels (zero shading)

PR = Performance Ratio: provides a coefficient for energy losses (see details below)

In the above equation, r is the yield of the solar panel, given in terms of the ratio of electrical power (in kWp) of one solar panel to the glass surface area (in m²) of the panel or module covering the solar cells. For example, the solar panel yield r of a PV module of 250 Wp with an area of 1.6 m² is 0.250 / 1.6 = 15.6 percent.

This nominal ratio is given for standard test conditions (STC): solar radiation equal to 1000 W/m², a cell temperature of 25°C, and a wind speed of 1 m/s. The nominal power of a photovoltaic panel under these standard test conditions is referred to as "watts peak" (Wp). All PV solar panels are rated in this way. For

example, a Wp rating of 250 indicates the theoretical, rather than the actual, amount of power that will be produced by a given PV panel, because Wp does not account for the losses that can occur. See "Performance Ratio" below.

H refers to the solar radiation or insolation value. You do not need to calculate this value yourself. You can find the solar radiation value for any location worldwide free of charge on several web sites, or from the above table for North America. For more extensive coverage of many other locations and countries, you can visit: http://www.photovoltaic-software.com/solar-radiation- database.php.

You'll have to find the global annual radiation incident on your PV panels, which is determined by your specific location (see tables above). Sometimes this needs to be modified with respect to your specific installation data, including the inclination (slope or tilt) and orientation or compass alignment (azimuth) of your panels. However, most of the major solar radiation websites will provide the solar radiation values for any given location, already taking into account the optimal inclination (number of degrees from the equator), and optimal orientation (i.e. due south).

PR (performance ratio): Knowing the PR is an important part of evaluating the quality of a photovoltaic installation, because it indicates how well the installation will perform independent of its orientation or inclination. It includes all losses. Factors that could cause energy losses that would affect the PR value include: the site; the PV technology; the type of inverter to be used; the sizing of the system; and the weather conditions determined by geographical location, as summarized below:

The following analysis of solar module performance factors will help explain the conversion from the solar module power rating (watts peak DC STC) to the energy (kilowatt-hours AC) produced by the PV solar system you propose to install at your home.

The "Performance Factors" for PV solar modules are summarized and quantified as follows:

Module Power Rating

Modules are rated in DC watts at STC by all manufacturers. For example: manufacturer rating 100 watts STC DC. All solar module manufacturers test the power of their solar modules under specific standard test conditions in the factory. The test results are used to rate the modules according to the tested power output. For example, a factory-test module that produces 100W of DC power would be rated and labeled as a 100W STC DC solar module.

Module Temperature Factor

The operating temperature of PV modules increases when the modules come in contact with the rays of the sun. As the operating temperature increases, the power output decreases slightly due to the properties of the solar cell conversion materials. This applies to all solar modules. The PV USA Test Condition (PTC) ratings, which are based primarily on the specific module temperature characteristics, take this into consideration. The PTC ratings are different for each module and can vary from approximately 87 percent to 92 percent of the STC rating. A typical decrease in power output is approximately 12 percent for crystalline-based solar modules. This decrease would result, for example, in a STC rated 100-watt DC solar module being PTC rated at approximately 88 watts DC.

Particulate Build-Up Factor

When a PV solar module is mounted on a rooftop, airborne particulates such as dust settle and accumulate over time on the glass surface of the module, just as dust settles on the glass windshield of your car wherever it may be parked. These particulates prevent a certain amount of light from reaching the module and therefore reduce the power produced by the module. As you know by now, modules produce more power when exposed to more light. Depending on localconditions and on the maintenance provided by the homeowner, the reduction in power from particulate build-up can be anywhere from 2 percent to 14 percent.

A typical value for this factor can be estimated at 7 percent, giving a particulate loss multiplier of 93 percent. A module installed in a wet weather climate wouldhave less "soiling" than a module installed in a drier climate, due to rainwater rinsing off the module's glass surface. Particulate build-up results in the power decreasing from 88 watts to approximately 82 watts (88W × 0.93 = 82W).

System Wiring And Module Output DifferenceFactor

Typical solar electric systems require that modules be connected to one another. The wires used to connect the modules create a slight resistance in the electrical flow, decreasing the total power output of the system, a phenomenon similar to what happens when low-pressure water flows through a long hose. In addition, slight differences in power output from module to module reduce the maximum power output available from each module. The system AC and DC wiring lossesand individual module power output differences could reduce the total system-rated energy output by 3 percent to 7 percent. A typical value for these losses is 5 percent. This results in the estimated power output decreasing from 82 watts DC to 78 watts DC (82 W × 0.95 = 78W).

Inverter Conversion Losses Factor

In order for the DC power from the solar modules to be converted to standard utility AC power, a power inverter (DC to AC) is installed along with the PV solar system. The conversion from DC power to AC power results in an energyloss of between approximately 6 percent and 8 percent. This is mainly due to energy losses in the form of heat, but this can vary between different inverters. Atypical value used for these losses is 6 percent. This gives a loss multiplier of

0.94. This results in the estimated power output decreasing from 78 watts DC to73 watts AC (78 W DC × 0.94 = 73 W AC). Now we have the power in alternating current after the inverter.

Module Tilt Angle Factor

First of all, not every roof has the optimal orientation or "angle of inclination" to take full advantage of the sun's energy. Non-tracking PV systems in the northern hemisphere should ideally point toward true south, although orientations that face in more easterly or westerly directions can work too, albeit by sacrificing varying amounts of efficiency.

Solar panels should also be inclined at an angle as close to the area's latitude as possible to absorb the maximum amount of energy year-round. A different orientation and/or inclination could be used if you want to maximize energy production for the morning or afternoon, and/or the summer or winter. Of course, the modules should never be shaded by nearby trees or buildings, no matter the time of day or the time of year. In a PV module, if even just one of its cells is shaded, power production can be significantly reduced (slightly more for central inverters or somewhat less if micro-inverters are used).

The module installation angle in relation to the sun affects the module energy output. The module produces more power (watts) and, as a result, energy (watt- hours), when the light source is perpendicular to the surface of the module. For this reason, solar module installations are often tilted towards the sun to maximize the amount and intensity of light exposure.

As the sun angle changes throughout the year (higher in the sky during summer and lower in the sky during winter), the amount of light falling directly on the module changes, as does the energy output. In southern California, a typical optimum tilt angle for average module power production over the course of a year in a fixed-tilt system is approximately 30 degrees. The typical southern California residential roof is tilted approximately 15 degrees. The reduction in the average annual energy output for a module, which is mounted at a south- facing, 15-degree tilt, is approximately 3 percent when compared to the optimal tilt angle of approximately 30 degrees.

This 3 percent reduction gives us a loss multiplier of 0.97 and results in a power reduction from 73 watts AC down to 71 watts AC (97 percent × 73 = 71).

For flat-mounted systems (zero tilt), the reduction in average annual energy output for a module is approximately 11 percent when compared to the optimal tilt of approximately 30 degrees.

Module Compass Direction/Azimuth Factor

The amount of sunlight shining on the module is partially dependent on the direction the PV system array is facing relative to the equator. As the sun moves across the sky throughout the day, from the east in the morning to the west in the afternoon, the compass direction (south, southwest, east, etc.) of the module affects the cumulative energy output. Assuming you're in the Northern Hemisphere, it's best to install a south-facing module in order to obtain the maximum amount of direct sunlight exposure on your PV modules. If the module is facing east or west, it will be exposed to less direct sunlight as the sun moves across the sky.

There is no loss factor for south-facing modules, so the estimated energy output (from one hour of exposure) for this particular example will remain at 71 watt hours AC.

If the module was not facing south, the estimated energy output would have been reduced. For example, the estimated energy output for a southwest-facing module would be reduced by approximately 3 percent.

Solar Irradiation Index Factor

The amount of sunshine on your modules is determined to a great degree by the extent of the sun's year-round intensity at your particular location, known as the solar irradiation index or solar insolation index.

Refer to the Solar Irradiance Map in Chapter 4, which illustrates and explains the different solar irradiance levels measured in average peak sun hours/day for any location in North America. The map accompanies the long table showing the "Sun Hours" for approximately 100 cities in the United States and Canada.

Every location on earth has a different amount of sunlight exposure throughout the year, referred to as solar irradiation, which is measured in kWh/M^2 expressed as "sun hours." For example, a coastal California city like Malibu or San Francisco will have a lower average amount of yearly sun hours compared with desert cities like Palm Springs or Phoenix because of fog and moisture in the air in coastal locations. Since solar modules produce power and energy when exposed to sunlight, the more sun hours a location receives, the more energy will be produced from a PV solar module installed at that location.

"One sun" refers to the peak noon sunlight power intensity in the middle of summer. "One sun hour" is the energy produced by the peak noon sunlight intensity in the middle of summer, over one hour. Because the sun's energy is converted by a solar module, recorded sun hour data for particular locations is used to help approximate the energy produced by a solar PV module.

The amount of sun hours for a particular location differs from day to day.

There are multiple sun hour data sources, which differ slightly from one another. The US Department of Energy and NASA have calculated average daily sun hour data for most locations for over 20 years, which helps predict yearly energy output. This recorded data shows an approximate daily sun hour average of 5.5 hours throughout the year for many southern California locations.

SUN HOURS

(Southern California = 5.5 daily sun hours), 391 watt hours AC per Day (71 watt hours AC × 5.5 h/day), equivalent to 142 kWh/year (391 Wh/day × 365 days divided by 1,000 W/kW).

A table in Chapter 4 specifies the average yearly sun hours available for more than 120 cities in the United States and Canada. This table also provides the average summer and winter sun hours, or irradiation index, for each city. In southern California, for example, there are approximately 7.1 sun hours per day during the summer and approximately 3.9 sun hours per day during the winter. These seasonal averages result in a yearly average of approximately 5.5 sun hours per day ((7.1 + 3.9) / 2 = 5.5).

Final Calculations of Solar Energy Output

In order to estimate the yearly energy production of a solar module, simply multiply the estimated module energy output (from exposure to one sun hour, 1000W/m2 over one hour), 71 watt hours AC (remember we used a PV panel of 100Wp adjusted by all the power-loss factors), by the number of sun hours for the particular location: 5.5 per day, in our example. This produces approximately 71W × 5.5h/day = 391 watt hours AC per day. Expressed in kilowatts, this is 0.391 kWh/day per panel. When estimating yearly energy production, the estimated daily energy production, .391 kWh AC, is multiplied by 365, the total number of days in a year. This results in approximately 142 kWh/yr AC energy production. Therefore, one 100-watt (100 Wp) DC module will produce approximately 142 kilowatt hours/year AC of electric energy under the specified conditions in this example.

If the module rating were higher than 100 Wp DC—for example, many polycrystalline modules have a rating of 250 Wp DC—you would simply multiply 142 by the ratio 250/100 (142 × 2.5) = 355 kWh/yr of AC power for one PV panel rated at 250 Wp.

The key figure to remember in all of this is the estimated module energy output, which is 0.71 watt hours AC for every 1.0 Wp DC of rated module power. This is equivalent to 29 percent total energy losses.

Once you make the commitment to go solar, the next step is to determine how big your solar PV system must be to meet the electricity needs of your home and to see if the total net cost is within your budget, taking into account any government financial incentives and subsidies, and then adjusting the size of your proposed system if necessary.

Start by reviewing your electricity bills over the past year to get an idea of your typical electricity usage measured in kilowatt hours. For example, in recent years the average American household used about 11,000 kilowatt hours of electricity per year, according to the US Energy Information Administration (EIA). Using the calculations above, to obtain 11,000 kWh of power over the course of one year in an average California city, we would need 31 modules rated at 250 Wp DC. The calculation is: 11,000 kWh/yr divided by 355 kWh/yr/panel = 30.9 panels (for PV panels rated at 250 Wp DC).

Of course, the above calculations were based on the solar irradiation index of an imaginary California city, and you will want to recalculate using the irradiation index in your particular location, which you can obtain from the table and map in Chapter 4.

Many utility companies also offer complimentary energy audits, and this can provide greater insight into your family's energy use habits and the basic requirements for your proposed PV solar system.

Some arid regions of the US Southwest can receive more than six hours of peak sun, while in the northeastern states it would be only about four hours. The process of comparing your power needs to your sunlight availability is known as your load calculation, and this simple calculation is critically important for planning the size of your PV solar system.

Let's take another, simpler example to understand how you may calculate the required size of your solar array. Let's use the typical North American family's monthly consumption figure of 900 kWh. To arrive at a daily consumption figure we divide the monthly figure by 30, the average number of days in one month. 900 kWh/month divided by 30 days/month = 30 kWh/day AC.

Now we want to calculate the size at which a solar array will produce 30kWh/day. We know this depends on peak sun hours available, and we'll use for this example five peak sun hours. Depending on your location, 30 kWh/day divided by five peak sun hours = 6 kW/day AC.

We need to remember to convert this back to DC power because all PV panels are rated in Wp DC. As detailed above, we learned the DC/AC conversion factor, allowing for all normal losses in 1.0 Wp DC = 0.71 W AC. Therefore, 6 kW/day AC divided by 0.71 = 8.45 kW/day DC.

Number of PV Panels of Different Ratings Required to Produce Specified Levels of Daily Power Consumption					
DAILY CONSUMPTION AC	EQUIVALENT DC POWER	NUMBER OF PANELS REQUIRED USING 200WP PANELS	NUMBER OF PANELS REQUIRED USING 240WP PANELS	NUMBER OF PANELS REQUIRED USING 285WP PANELS
4,000 W/d AC	5,634 W/dDC	28	24	20
5,000 W/d AC	7,042 W/dDC	35	29	25
6,000 W/d AC	8,450 W/dDC	42	35	30
7,000 W/d AC	9,859 W/dDC	49	41	35
8,000 W/d AC	11,267 W/dDC	56	47	40

Note: The number of panels is rounded off to the nearest whole number.

Therefore, for a 6,000 W/d AC (8,450 W/d DC), you would need forty-two 200Wp panels (8,450 200 = 42), or *thirty-five 240Wp panels (8,450 240 = 35)*, or thirty 285Wp panels (8,450 / 285 = 30). This calculation is included in the table below.

Those of you with three-or four-bedroom homes who are putting in solar PV systems will want to offset most of your electrical requirements with a PV solar array that is between 4kW/d and 8kW/d (daily consumption AC). Using this simple formula provided above, we can calculate the number of panels for different power levels as described below

If your monthly average consumption figure varies much from the 900 kWh/month example, or if you receive fewer than five peak sun hours on average, then it's simple math to calculate your own daily consumption figure and the number of panels in accordance with their rated capacity. The basic formula is as follows:

Monthly consumption___kWh/month, divided by (30 d/m × 1,000 W/kW) =_____W/d, divided by number of peak sun hours_____h/d, divided by the capacity of each solar panel _____Wp DC =___, the total number of panels required for your system. (Note: To be practical, we round up or down to the nearest whole number.)

Links to Online Energy Output and System Size Calculators for PV Grid-Connected Systems

The National Renewable Energy Laboratory (NREL) developed the world's most widely used computer model to estimate the energy production and energy cost of grid-connected photovoltaic solar energy systems. It can be used for existing or proposed PV solar systems anywhere in the world. This service, and specifically their PVWatts calculator, enables homeowners, small building owners, installers, and others to quickly develop estimates of performance for potential PV solar installations. The calculator can be accessed here: http://pvwatts.nrel.gov

In the UK, Scotland, and Wales, there's a quality solar energy calculator that estimates the income and savings the homeowner can receive from the domestic feed-in tariff scheme, which is available for eligible PV installations of up to 4kWp. The calculator uses the latest tariff rates and is available here: www.energysavingtrust.org.uk/domestic/solarenergy-calculator.

A site for an easy-to-use online solar energy calculator for US and Canadian PV solar customers that computes how many tons of CO_2 emissions any proposed PV solar rooftop system will avoid can be reached through this web address: www.solarenergy.org/solar-calculator.

The following website is for the USA-Canada market, and it provides some useful programs that can calculate the size of the PV solar rooftop you need, as well as system costs. It also helps you determine the government financial incentives you may qualify for.

Here is a link to the calculator in reference:

www.solar-estimate.org/?page=solar-calculator

The following links to solar calculators each enable you to build your own customized calculator to match your particular desired conditions or energy consumption level:

www.findsolar.com/Content/SolarCalculator.aspx (USA–Canada)
www.energymatters.com.au/climate-data (Australia)

www.affordable-solar.com/residential-solar-home/Residential-Calculator (USA)

The last website above also provides a national map of peak sun hours summarized by state.

Chapter 10 contains information about numerous websites that provide reputable cost savings calculators, which can automatically compute the annual cost savings of your proposed PV solar rooftop system as compared to your future yearly electricity bills if you don't install a solar system.

Finally, it's important to note that the figures generated by online solar calculators should only be used as a rough estimate. Some calculators may not include data on federal, state, or local solar incentives and rebates. Site-specific factors may also influence designs and output that an online calculator would not factor into the computations. A solar contractor will usually be able to provide you with a more accurate estimate.

Chapter 6 - Government Incentives, Rebates, Subsidies, Grants, Tax Credits, and Private Leasing Programs

Government economic incentives will lighten the financial burden. Now that your system is planned, you only need to figure out the financing alternatives and sources for your solar power system. There are also some standard legal issues, mainly to do with permits.

If you live in the United States, we recommend you visit a government organization known as the Database of State Incentives for Renewables & Efficiency (DSIRE) at www.dsireusa.org. This website has an up-to-date database of energy efficiency financial incentives. The United States government currently provides a 30 percent tax credit for eligible costs of your solar array, and you can find more grants, loans, and tax credits by consulting your state or local government. DSIRE also has information on utilities that will buy electricity back from you.

While you're reviewing this material, keep in mind that some incentive programs are exclusive to PV solar systems installed by a certified installer. These aren't available to DIYers who install their system themselves, so be clear on the eligibility requirements of these programs.

Now all you need to do is select a good solar energy installer. Even if you're a handyman type, make sure to hire a licensed electrician for a safety inspection and for helping with wiring challenges. Electricians are licensed and can pull permits with your local building office. In principle, you can install your system all the way into your breaker box, but it will need inspection. In many jurisdictions, you aren't allowed to install a solar energy system without obtaining the permits required by your local government authority. Local laws may require safety measures, including building permits and compliance with national electrical codes, so be sure these requirements are met before you begin to install your PV system.

A very typical PV solar rooftop installation of (2 rows × 6 panels) = 12 poly panels that would qualify for optimal government incentives in most states of the United States.

Quite often, the greatest challenge is the inability of homeowners to obtain financing for the often-hefty down payment on equipment and installation. In an effort to encourage the adoption of renewable energy systems and support the growth of the solar industry, federal, state, and many municipal governments, as well as some utility companies, offer cash rebates and/or other economic incentives to subsidize the cost of PV solar system installations.

The various incentives can be confusing at the beginning. In general, the most valuable economic incentives are federal tax credits and state and local rebates. I'll summarize each of these programs below, and we'll also discuss lesser-known incentives that may be valuable, especially as more states look for alternatives to upfront rebates.

Many states, municipalities, and utilities offer rebate programs or cash incentives. These rebates usually take the form of a per-watt cash rebate ranging from $1.50/watt to $5.00/watt. For example, a rebate for a 5kW system under a program offering $3/watt would equal $15,000. These cash rebates can offset the cost of a solar installation by as much as 45 percent. Most cash rebate programs have an upper limit to the refund, often set at 5kW, and PV systems bigger than that can only claim a rebate up to the specified limit.

In most locations, as the state's renewable-energy targets are achieved, the available per-watt rebate ceiling may be reduced. That means a state may offer a $4/watt rebate today, but only a $2.50/watt rebate down the road, once a certain number of solar installations have been achieved statewide.

Normally, a finite amount of money is allocated for rebate programs. Due to high demand, the funds are often quickly exhausted. However, once the funds are gone, many programs offer waitlists until the next round of funds is made available. In addition to rebates, tax credits may also be available. At both the federal and state level, laws have been put in place to credit a percentage of the purchase price of a PV solar power system against a solar energy system owner's annual tax bill. A tax credit is more valuable to the taxpayer than an equivalent tax deduction. That's because a credit reduces your taxes dollar-for-dollar, while a deduction only lowers your taxable income before applying the corresponding tax rate that gives your "tax payable" amount.

At the federal level, the Federal Investment Tax Credit entitles owners of both commercial and residential renewable-energy systems to a credit of 30 percent of the net system cost, with no set limit, for all systems placed in service before December 31, 2016. The credit can be carried forward 15 years or back three years. It's unknown if this deadline for federal solar investment tax credits will be extended. A number of states are also offering tax credits equaling a percentage of the installed cost of any PV solar energy system. The percentage varies from state-to-state.

I should note that if a customer receives a rebate, any tax credit is computed on the remaining balance, not the pre-credit total. For example, if the total solar energy system cost is $35,000, and a customer receives a $15,000 rebate, the credit of 30 percent is calculated only on the remaining system cost balance of $20,000, for a total credit of $6,000.

Most states and some municipalities can also provide some sort of property tax exemption for individuals and businesses that install PV solar energy systems. Even though a solar installation will normally increase your property value by tens of thousands of dollars, states that offer a property tax exemption do not increase your tax liability to match your property's new value. This is a nice plus for the homeowner.

Some states offer a 100 percent exemption for the entire value of the system, while others offer a partial exemption (typically 75 percent or some other percentage of the total system's value). Depending on the state, the exemption may apply for one of the following options: 1) the first year of a system's installation; 2) a pre-set term that typically ranges from 10 to 20 years; or 3) for the lifetime of the solar energy system (no time limit). Usually, one of the last two options will apply.

Other states provide an exemption only for the added value of the renewable energy system above the value of a conventional energy system. For example, if a solar energy system costs $30,000 to install and a conventional system costs $18,000, the property tax will be assessed on an increased value of only $12,000.

When filing taxes, a customer may only be able to take full advantage of the exemption if the system was installed and operative throughout the 12-month period preceding December 31st of the tax year. If the system was operative for only some portion of the 12-month period, the exemption will likely be reduced proportionally.

Along with property tax exemption, many states don't collect state sales tax on the cost of renewable energy equipment, which can significantly reduce the upfront costs of a solar installation.

Renewable Energy Credits, or RECs, are certificates issued by the state for the amount of clean solar energy that a grid-tied solar system produces. In some markets, RECs may also be called SRECs (Solar Renewable Energy Credits/Certificates) or Green Tags. Utility companies may purchase RECs from clean power generators as one way to meet the Renewable Portfolio Standards (RPSs) that have been put in place by state legislatures to mandate that utility companies obtain a percentage of their power from renewable sources. In addition, a company such as a manufacturer or coal plant may also buy RECs on the open market as a way to offset the greenhouse gases they emit. The value of a REC on the local market varies by state, and it fluctuates based on supply and demand. The means by which RECs can be sold also vary by state.

As you now know, net metering is the process some utility companies use to keep track of any extra power that a grid-tied solar system produces as well as the amount of power that is fed back into the grid. During the summer months, or during daylight hours, a household or business uses more electricity than during winter or at night. On a hot summer day, therefore, a household PV solar system will likely produce more energy than is required by the home, and on a cold winter night it's likely to use less. In a net metering arrangement, the utility keeps track of energy usage and stores any extra power your solar system generates. During those times when a solar energy system generates more power than is being used, the extra electricity makes the electric meter spin backwards. At the end of the month or the year (depending on the utility company), the customer receives a bill that reflects the net result of what the solar panels produced against the amount of electricity the household or business consumed. With net metering, the annual electricity bill could be, for some installations, as little as $100. In some states, a utility company will not pay the customer if the solar energy system generates more power than is used during the course of a year. In this case, the annual bill will be $0, but all excess power will be donated to the utility company. In other states, utilities will pay for the extra production at various rates. To optimally take advantage of net metering, a solar system should be sized to meet your home's power requirements in periods or hours of maximum demand.

Having jumpstarted the solar industry in countries like Germany and Spain, feed-in tariff (FiT) schemes were introduced in the US, with states like Vermont and California leading the way. A FiT is an incentive system in which the utility company compensates a person who is generating clean power. Typically the FiT agreement covers a time period of up to 20 years. Under some FiT schemes, a residence or business that installs a grid-tied PV solar system essentially becomes a mini-utility that generates power both for the home and for the grid. Some months the utility company may send the system owner a check that may even be higher than the equivalent amount of the electric bill for that month, or the utility may credit the account each month under the net metering system up to a maximum of the total year's power consumption of the PV system. Some state FiT programs are more short-term and serve as a quick way to increase solar installations and solar power generation. FiT programs in some jurisdictions may be reserved for large-scale installations, and they may exclude residential installations.

For homeowners who haven't saved quite enough money to get started, taking out a loan might be a viable option. Several states and utility companies are leading the way with innovative programs that provide low-interest loans to homeowners and businesses for PV solar installations. In New Jersey, for example, utility company PSE&G recently committed to provide more than $100 million toward the financing of solar system installations over a two-year period. New Jersey and many other states are firmly committed to financially assisting homeowners who wish to invest in a PV solar rooftop system for their home. In California, cities and counties are making loans available to consumers who are then able to repay the loans as part of their property taxes (this scheme is often referred to as the "Berkeley Model" after the city that introduced it: Berkeley, California). Loan balances are transferred to whoever owns the property if it's sold during the course of the loan repayment period.

Before you make the commitment towards a loan, a grant may be available in your area to help fund your project. Unlike loans, grants do not have to be repaid. Some states offer grant funding for residential and commercial properties for the installation of PV solar systems. Typically, grants are awarded on an agreed-upon dollar value per watt up to a maximum, e.g., $1.25/watt up to $10,000.

With so many different types of incentives available and so much variation state-to-state, it's essential for anyone contemplating a solar installation to do his or her homework to ensure that all possible benefits are realized.

Under various schemes, including the California Solar Initiative, individuals and companies can receive cash rebates for electricity provided to the grid from PV systems that are owned outright.

In the case of leased systems, residents of buildings with rooftop PV installed pay a third party leasing company to provide and install the system. The leasing company is also responsible for maintenance. The appeal for a householder of leasing a rooftop PV system is the simplicity of becoming involved in solar power generation and saving money on energy bills without facing the immediate capital outlay associated with installing and running a system by themselves. The benefits to a lease company include receiving the 30 percent tax credit (as long as the program stays in force) as well as receiving renewable energy credits against the amount of electricity generated and carbondioxide offset of the leased PV solar system.

According to a recent CPI (Climate Policy Initiative) study, the number of leased rooftop PV installations in California compared to the number of total new solar installations grew from 10 percent of all new installations in 2007 to 56 percent of all new installations in 2011, jumping to 75 percent of all new installations in 2012. At present, leasing is allowed in most, but not all, states. In 2015, the state legislatures of Georgia and of South Carolina approved legislation allowing third-party ownership of PV solar systems. Florida and North Carolina are expected to approve similar legislation in late 2016 or 2017.

The solar leasing market in the United States received another positive boost recently with the announcement that global solar panel manufacturer SunPower and Bank of America Merrill Lynch have joined to provide the rooftop PV solar sector with $220 million in financing, which will assist US homeowners in financing solar power systems through solar leases provided by SunPower. An estimated 50,000 US households have already signed lease agreements under this program, which offers low monthly payments and includes one of the few direct-from-manufacturer performance guarantees for installers.

SolarCity, SunPower, Sungevity, and Solar3D are only a few of the major US PV solar system manufacturers and suppliers of rooftop systems offering leasing packages to homeowners. Of these companies, SolarCity is considered to be the most aggressive group in the PV solar system leasing sector, but SunPower is also very competitive, and the others are expanding their solar system leasing businesses.

In Europe, leasing PV solar rooftop systems has not yet been established because they do not have net metering systems. However, there are a few companies in Germany (including DZ-4) that are developing different business models for PV solar leasing, and many of these companies have initiated pilot projects. Likewise, several large German banks are studying leasing models for PV solar residential systems. These systems in Germany will normally incorporate a PV battery bank storage system. Feed-in tariff programs, on the other hand, have had amazing success in helping solar customers to invest in PV solar systems in Germany and throughout Europe.

Chapter 7 - Environmental Benefits of PV Solar Systems

Many prospective buyers of a rooftop PV solar system ask: How much harmful CO_2 will we avoid when we go solar? Reliable statistics are available that show how much conventional carbon-based fuel is consumed in the production of a given amount of electricity, measured in kWh. Also, we know how many tons of CO_2 are emitted by the consumption of carbon-based fuels to generate a given amount of electricity, *i.e.* for every 1 kWh or 1,000 kWh (1 MWh) of power. Therefore, if we know the amount of power consumed by a PV solar system in kilowatt hours or megawatt hours, we can calculate exactly how many tons of CO_2 gas emissions are prevented from entering the atmosphere by virtue of using photovoltaics renewable energy.

To produce 1,000 kWh of electricity, a conventional power plant using carbon-based fuels will produce 0.706 metric tons of CO_2, on average. A photovoltaic solar system will produce the same 1,000 kWh of electricity with zero CO_2 emissions. Therefore, if we know the average daily or monthly electric consumption of a residential PV rooftop system, we can calculate the equivalent of CO_2 emissions that are prevented from entering the atmosphere over a month, a year, 20 years, or a period of any length.

A photovoltaic residential rooftop system of average size may have an installed capacity of 7 kW/h. If this system receives an average of five hours of sunlight per day, this would produce approximately 7kW/h × 5 h/day = 35 kWh/day × 365 days/year = 12,775 kWh/year. Using the above figure of 0.706 metric tons of CO_2 per 1,000 kWh and multiplying that times 12.775 kWh/year would be equivalent to 9.02 metric tons of CO_2 emissions that were prevented from entering the atmosphere over the course of 12 months.

If we take the useful life of the PV system as 25 years, then this system would save the equivalent of between 225 and 270 metric tons of CO_2. When we multiply this by the millions of systems installed globally, it represents a significant contribution to the reduction of global warming.

To contribute this much to saving the environment while at the same time saving money with every monthly electric utility bill is very meaningful to most homeowners. It's a true win-win situation.

Chapter 8 - Maximize the Benefits from Your PV Solar EnergyRooftop Installation

There can be huge benefits and savings if you install energy-efficient lights and appliances throughout your home before installing solar energy. By doing this you willsave money and benefit in two distinct ways.

First, if you're seriously looking at using solar energy in your home, you shouldknow that by replacing existing energy-inefficient systems and appliances with newer, more efficient ones, and by being more conscientious with your electric consumption habits, you can substantially reduce your monthly electricity bills

—often by as much as 40 percent. This, in turn, can reduce the size and overallcapital cost of any PV solar energy system to be installed in the home by a similar percentage. Conservation will also make the investment in solar energymore effective and reduce your payback period.

You can draw up a useful electrical energy efficiency checklist, as follows:

Refrigerators

Most homes have at least one or two refrigerators. If your model has a manufacture date (found on the name plate on the back of the unit) prior to 2002,chances are it's one of the older, inefficient models, probably consuming about500 to 750 watts per hour. These units should be traded in for a newer, quieter,more energy efficient model with a rating of 350 watts.

Newer models have additional advantages and new convenience features, such as automatic ice cube makers, auto defrost, useful interior designs, and attractive exterior designs with longer-lasting, more user-friendly door handles.Eventually, the monthly savings from the electric bills will pay for the cost of the new refrigerator.

Air Conditioners

In many parts of North America and Europe, and in all tropical countries, homesuse unsightly window units. These are often older, noisy, and energy inefficient.If not maintained and cleaned properly, some older units will leak water insidethe house—a truly annoying situation. By universal convention and industry standards, the power and electric consumption of air conditioners is indicated bythe BTU (British Thermal Units) rating. A very small home may use a unit as small as 1,200 BTUs. In larger homes with large rooms, there may be several units as large as 36,000 BTUs. The earlier models—dating from eight to twelve years ago—tend to be inefficient and consume a lot of electricity.

If these older units are recycled and replaced with modern, high-efficiency ceiling units, the resultant savings to the monthly electric bills can be substantial. In addition, large open cracks or spaces around outside windows and doors can result in a lot of unnecessary losses and are easy to fix with rubber or foam-based door-and window-sealer strips.

Turning off the air conditioner when you leave the room can also reduce consumption. Even more savings can be achieved by switching from the air conditioner to a floor or wall fan at night, when the outside ambient air temperature goes down. Combining these recommendations can make for big reductions to the monthly electric bills, and the savings can be seen within a month or two of making the changes. If you're designing a new home or doing a major renovation, and if the size of your home is large enough, then a modern, high-efficiency central air conditioning unit can be considered. These offer benefits including low electric consumption, even temperature control, less maintenance, and improved aesthetics.

ENERGY CONSUMPTION CALCULATION

The energy consumption "E" in kilowatt-hours (kWh) per day is equal to the power "P" in watts (W) times the number of usage hours per day "t" divided by 1000 watts per kilowatt:

$$E_{(kWh/day)} = P(W) \times t(h/day) / 1000(W/kW)$$

Electric Water Heaters, Electric Clothes Dryers, and Portable Electric Space Heaters

The older models of these appliances are generally not efficient, and their combined electricity consumption often comprises a sizeable portion of the monthly electric bill. Typical power consumption ratings are usually in the following ranges:

Electric Water Heaters: 4,000 to 5,000 watts Electric Clothes Dryers: 3,000 to 4,000 watts Electric Space Heaters: 2,000 to 2,500 watts

Adopting the formula for energy consumption calculation on the previous page, and using the lower end of the range for each appliance and a reasonable average household usage time, we get the following energy consumption levels:

- Electric Water Heaters: 4,000 W × 3 / 1000 = *12 kWh*/day = 4,380 kWh/yr
- Electric Clothes Dryers: 3,000 W × 2.5 / 1000 = *7.5 kWh*/day = 2,737.5 kWh/yr

- Electric Space Heaters: 2,000 W × 8 *1000 = 16 kWh*day = 5,840 kWh/yrTotal = 12,957.5 kWh/yr

If you pay $0.12/kWh for your electricity, this represents an annual cost of $1,555 per year.

Substantial savings can be achieved by recycling these electric appliances and converting to natural-gas appliances or, where natural gas lines are not available, using large residential propane tanks with small diameter copper tubing.

It is highly recommended to study the advantages of installing a solar rooftop water heating system instead of using traditional water heaters. Solar thermal water heating systems are the most efficient of all renewable energy systems.

Their energy-conversion efficiency (sunlight to heat for hot water) is approximately 75 percent. Compare this to PV solar panels (converting sunlight into electricity) that typically have efficiency factors in the range of 18 percent to 21 percent. As a result, a solar thermal rooftop water-heating system has a very short payback period and will substantially reduce the size of the PV solar system needed for the same residence. The two systems are very compatible, assuming sufficient space is available for both. Usually only three or four hot water collector panels are sufficient for the average home.

Home Lighting Fixtures

Incandescent lighting uses those standard old-fashioned round light bulbs that burn your fingers and have to be replaced frequently. Incandescent lighting is still in widespread use around the world and is an extremely inefficient technology. These fixtures typically convert up to 86 percent to 92 percent of the electric input into heat (a huge loss) and only 8 percent to 14 percent into light.

This is very wasteful and easy to remedy by replacing these incandescent fixtures with compact fluorescent lamps, commonly referred to as "CFLs," or—even better—with "Light Emitting Diodes" (LED lighting).

Incandescent;

Compact Fluorescent;

LED.

The following chart illustrates the striking differences between the old and the new lighting technologies.

As the table indicates, LED lighting uses far less power (watts) per unit of light generated (lumens). Traditional incandescent bulbs use nine to twelve times more power than LED lights for the same light intensity or lumens rating. LED lighting also substantially lowers electric bills. But there are two other economic advantages. One advantage is that LED lights have an immensely longer useful life than incandescent bulbs and will need replacing far less often, saving the homeowner money and a lot of bother buying new bulbs and changing burnt-out ones. A final advantage: By converting to LED lighting, homeowners who want to install a PV solar rooftop system will need fewer solar panels to power the home's lighting.

When you add up the power savings by using LED lighting and follow the other energy saving recommendations above, you'll not only save money every month by consuming less electricity, you'll also reduce the size of the required PV solar rooftop installation, once again saving a lot of money. And you'll get to enjoy those stylish, attractive new appliances!

There is one other significant advantage to LEDs that is mainly environmental but also has long-term economic benefits for society and the planet as a whole. When you multiply the LED reductions in kWh/year by hundreds of thousands of homes, you'll see that LED lighting can help to reduce greenhouse gas emissions from power plants due to a corresponding decrease in the carbon-based fossil fuels needed to power the residences that convert from incandescent lighting to LED lighting.

Lighting: Energy Efficiency Cost Comparison Chart			
ENERGY EFFICIENCY AND ENERGY COSTS	LIGHT EMITTING DIODES (LEDS)	INCANDESCENT LIGHT BULBS	COMPACT FLUORESCENTS (CFLS)
Life Span (average) Note: Incandescent bulbs have terribly inconsistent quality and can often burn out in 800 hours or less.	50,000 hours	1,200 hours	8,000 hours
Watts of electricity used (LED and CFL equivalent to a 60-watt incandescent bulb).	6-8 watts	60 watts	13-15 watts
Kilowatts of electricity used (30 incandescent bulbs per year or equivalent)	329 kWh/year	3285 kWh/year	767 kWh/year
ANNUAL OPERATING COSTS FOR 30 INCANDESCENT BULBS PER YEAR OR EQUIVALENT			
USA (US$0.12/kWh)	$39.48/year	$394.20/year	$92.04/year
France (US$0.19/kWh)	$62.51/Yr	$624.15/Yr	$145.73/year
UK (US$0.20/kWh)	$65.80/Yr	$657.00/Yr	$153.40/year
Japan (US$0.26/kWh)	$85.54/Yr	$854.10/Yr	$199.42/year
Australia (US$0.29/kWh)	$95.41/Yr	$952.65/Yr	$222.43/year
Spain (US$0.30/kWh)	$98.70/Yr	$985.50/Yr	$230.10/year
Germany (US$0.35/kWh)	$115.15/Yr	$1,149.75/Yr	$268.45/year

Denmark (US$0.41/kWh)	$134.89/Yr	$1,346.85/Yr	$314.47/year
ENVIRONMENTAL IMPACT			
Contains toxic mercury	No	No	Yes — mercury is very toxic to human health and the environment
Compliant with International Electric Codes and Standards	Yes	Yes	No — contains 1mg-5mg of mercury andis a major risk to the environment
Carbon dioxide emissions (30 bulbs per year)	451 pounds/year (205 kgs)	4500 pounds/year (2,045 kgs)	1051 pounds/year (477.7 kgs)
OTHER IMPORTANT CHARACTERISTICS			
Sensitive to high and low temperatures	No	Somewhat	Yes — may not work under negative 10 degrees Fahrenheit or over 120 degrees Fahrenheit
Sensitive to humidity	No	Somewhat	Yes
Effects of on/off cycling (Switching a CFL on/off quickly, in a closet for instance, may decrease the lifespan of the bulb.)	None	Some	Can reduce lifespan significantly
Light turns on instantly	Yes	Yes	No — takes time to warm up

Durability	Very durable. LEDs can handle jarring and bumping.	Short durability. The glass bulb or the filament can break easily.	Not very durable. Glass can break easily.
Heat emitted	3.4 BTUs/hour	85 BTUs/hour	30 BTUs/hour
Failure modes	Not typical	Some	Yes — may catch on fire, smoke, or omit an odor

LIGHT OUTPUT			
Lumens	Watts	Watts	Watts
450	4-5	40	9-13
800	6-8	60	13-15
1,100	9-13	75	18-25
1,600	16-20	100	23-30
2,600	25-28	150	30-55

In the case of the United States, where electricity costs average about

$0.12/kWh, the conversion of 60 incandescent 60W bulbs (or 36 100W bulbs) toLED lighting alone could save the residents of a single home about $722 per year plus the savings in purchases of replacements for burnt-out bulbs—let's say

$150/year—for an overall per-home savings of $872/year. If we multiply this bythe minimum life of a PV solar rooftop system—25 years—this would be a totalsavings of $21,800 and this is only for the lighting!

T-12 fluorescent lamps (the old 1 1/2-inch diameter tubes)

This is really old technology—from the 1940s and '50s—but there are still lots of these lighting fixtures around. They should be replaced by the newer generation of T-8 (1-inch diameter) and T-5 (5/8-inch diameter) fluorescent fixtures, which use a more efficient ballast design and have a higher operating frequency. They're more energy efficient, less obtrusive, and more modern looking, and they give off better light by using an improved coating on the inside of the tube. Care must be taken to recycle or dispose properly of the old fixtures and T-12 lamps.

Automatic Light Controllers and Dimmers

As lights are so easily left on when not in use, it's wise to install timer switches or occupancy sensors to overcome the "forget-to-switch-off" syndrome. After a light is switched on, a timer switch will automatically turn off the light after a set time interval. The best applications are for areas where the light is needed for a short time only, such as in closets, hallways, stairwells, and garages.

Occupancy sensor light controllers are available in two variations: ultrasonic sensors and passive infrared sensors. Occupancy sensor lights receive a signal whenever a person moves into the predetermined conical space, set at a convenient diameter where the light strikes the floor. As long as there is someone moving in this space, the light will remain on. When there is no motion, the occupancy sensor light will turn off the light following a predetermined time interval that can be preset by the customer, usually with four time interval options. These controllers are suitable in kitchens, laundry areas, garages, workshops, and porches. They aren't recommended where there's a lot of movement by pets. Also, they aren't convenient where a person might be immobile for relatively long periods, as the inactivity will cause the sensor to turn off the light. Ultrasonic or dual-sensor type controllers are often used in bathrooms and bedrooms.

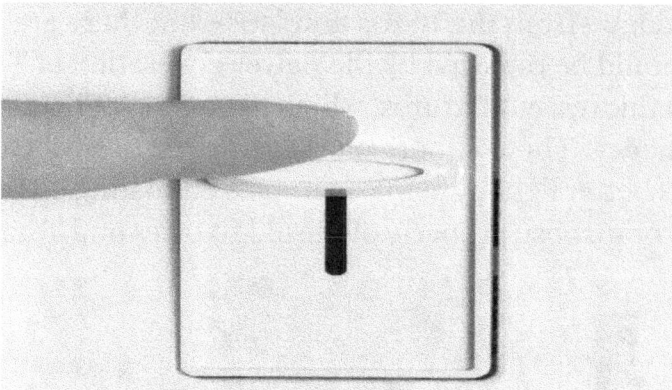

The original generation of CFLs could generally not be used in conjunction with dimmers, but the new generation of CFLs has overcome this limitation.

Dimmers can save electricity as well as creating a calm atmosphere or romantic mood, when desired.

Double-Pane Windows

If you're designing a new home or doing major renovations to an older one, you'll want to consider installing double-pane windows throughout the house. For both heating and cooling, the extra insulation provided by these windows is a good investment and can produce substantial energy savings, which will positively affect your monthly electric bills while providing more even heat control and more effective and comfortable cooling.

Insulation

Many homes are built with improper or insufficient insulation, especially older homes. By having an insulation contractor inject granular insulation in the spaces between the inside and outside walls, you can substantially reduce your heating and cooling costs. These spaces are easily accessible to the contractor through the attic. The attic itself should also be inspected, to ensure that it's properly insulated with fiberglass packs that fit between the joists of the ceilings.

Weather Stripping

Installing weather stripping is an overlooked but simple way to prevent or reduce heat and cooling losses that might be costing you a lot more than you would guess. Oftentimes, the weather stripping on the bottom of a door gets damaged or torn off and not replaced.

Location and Use of Windows for Optimal Natural Ventilation

If you're building or planning to build a house, study the location of all the windows—maybe in consultation with your architect or building contractor—to see if you can relocate any windows to achieve better natural ventilation in the summer months.

Some other easy energy saving recommendations to consider include installing a low-flow shower-head and taking shorter showers to conserve both electricity and water. You can also lower the temperature setting of your water heater while waiting to replace it with a solar thermal water heating system and insulate all your exposed hot water pipes to reduce heat losses. Motivate your family to commit to being conscientious about turning off lights, appliances, and electronics when they're not in use. After you and your family put these commitments to practice, look over your monthly electric bills together to see the positive results you've all been working toward.

By incorporating these recommendations, you'll benefit in the short-term by reducing the size of the PV solar rooftop system you install as well as the capital investment needed to install it. You'll also benefit in the long-term through savings on monthly electric bills and winter heating bills.

Chapter 9 - Where to Find Professionally Certified Solar Installers in Your Area

If you're planning to install a PV solar rooftop system for your home, you'll want to know what your options are. If you've already decided to hire a solar systems installer-contractor, or even just a "solar system coach," then you can go online and visit the NABCEP website (details below).

First and foremost, double-check that the prospective solar PV installer has received training and a professional certification to prove their knowledge and competence. Many states require solar contractors to meet some type of licensing requirements. If your state isn't one of those, you'll need to confirm your contractor's credentials yourself to ensure a job well done. The most acceptable seal of approval comes from the North American Board of Certified Energy Professionals (NABCEP), which now certifies solar PV installers across the USA.

You can use the NABCEP website (www.nabcep.org/certified-installer-locator) to find professionals in your area who have received their certification. Many states now require that you use an NABCEP-certified installer in order to participate in any incentive programs. Here's the website of the NABCEP, where you can locate solar installers in most cities in the USA: www.nabcep.org/certification/pv-installer-certification.

If you're not able to find an installer with an NABCEP certification, you can look for other certifications to guarantee the professional installation of your solar electric system. Many states require that a licensed electrician be on your installation team. Not all licensed electricians are familiar with PV systems, so ask if they have received solar-specific training and/or possess experience installing solar systems. Any good professional installer or electrician should be happy to provide you with references from a couple previous customers. While you're at it, you might ask whether the potential installer has received formal training installing the particular brand of solar equipment you plan to buy. Many manufacturers offer continuing education to solar installers to ensure that they're familiar with all the features and intricacies of their unique systems and updates to their technology. Here are some guiding questions you can ask when contacting the provided references:

- How long did it take to complete the installation, from initial site visit through to the completion of the project?
- Were there any challenges involved in the installation of the solar system? If so, how did the installer cope with these challenges?
- How has the system performed since it was installed? Has it met expectations for efficiency and good service?
- How have any requests for maintenance been handled by the installer? Did it take long for the installer to respond and was the maintenance service satisfactory?

Solar Power World is one of the solar industry's leading sources of information for solar technology, developments, and installation news. This organization recently published a list of the leading 250 solar contractors in theUSA. The list ranks companies according to the volume of PV systems they have installed in the residential and commercial solar energy markets.

Residential installations are listed separately. This resource may provide you with some useful contacts: www.solarpowerworldonline.com/top-250-solar- contractors.

Your state will have a contractor's licensing board, through which you can verify the credentials of the potential solar PV installer. I strongly recommend you contact them before signing any contracts, as the board should be able to alert you to any serious complaints received against a specific installer. Finally,be sure to verify the presence of sufficient liability insurance by asking to seeyour contractor's insurance papers.

Having an installer who is up-to-date on all of the latest advances in the ever-evolving world of solar photovoltaics is a major plus. The PV solar industry seesnew technologies for improving efficiencies and boosting performance on almost a monthly basis. The more current your installer's knowledge is, the morelikely he'll be able to provide the highest-efficiency system for your unique situation, while averting any potential problems.

In many cases, experience will be just as important as formal training. Since structured certification and training programs are just now becoming conventional, there are many installers who have been apprenticed into their current positions, and their experience can be highly valuable. Here are some of the queries you should consider to see whether your prospective installer's company is well-seasoned:

- Does the company have experience with both grid-tied and off-gridsystems?
- How long has the company been in business, and how many solar energyinstallations have they completed? Installers who have completed a largenumber of installations over many years will have ample experience to ensure you get good results.
- Determine whether the company has both commercial and residentialexperience.
- Ask the installer to describe their last installation. Was it recent or sometime ago?
- What types of add-on technologies have the company installed to improvemonitoring and system efficiency?

Long-term customer service is of paramount importance when choosing a solar installer. Your final query for any solar installer should have to do with how they will treat you over the long term. In all likelihood, you should considerchoosing an installer with warranties and service agreements that will make yoursolar installation a hassle-free investment over the life of the system.

If your solar products don't come with their own warranties from the manufacturer, you should ask whether the installer will warranty the equipment.Whether the warranty comes from the manufacturer or the installer, you should try to obtain coverage for the greatest number of conditions over the longest possible time to make sure you're not left with any bills for repairs to a non- functioning or defective system.

Figuring out what exactly is included in the service agreement offered by asolar installer can be tricky, especially if the agreement is full of legal terminology and technical information. Here are a few contractual questions toask your proposed solar contractor:

- For how many years will service be provided after the installation?
- Are the installers certified to provide maintenance to a PV solar rooftopsystem?
- Will you receive any training from the installer on how to maintain yoursolar system for maximum energy production?
- Will a yearly check-up be performed on your system to determine whetherit's performing as expected?
- Are web-based monitoring systems available to track the system'sperformance?
- What types of repairs and replacements does the service agreement cover?
- Does the service agreement state a guaranteed maximum length of timebetween a service request and the service call?
- What happens when the equipment doesn't perform as efficiently as promised? Are adjustments to improve efficiency included in the service agreement?
- Once the service agreement expires, how much will the installer charge youfor repairs or service? Is it based on the cost of the new parts used plus anhourly rate?

As with all other inquiries during your evaluation of potential installers, the answers you receive from any major equipment vendors (for PV modules, inverters, etc.) will also give you a clear idea of how you'll be treated as a customer over the long-term. Once you've asked all of your questions and determined that you've found the right installer, get the installer's commitments in writing so you receive exactly the service you've been promised.

Lastly, if a relatively large photovoltaic system will be installed on your roof,you might want to hire a licensed roofer to calculate the stress the proposed new system will place on your home's structure. It's important to determine whether the load will be within safety tolerances. The size and type of your home PV system will determine the weight of the installation and will play a significant role in determining how to safely and securely fasten the equipment and support hardware to the roof surface.

Chapter 10 - Photographic Step-by-Step Guide on How to Install aComplete PV Solar Rooftop System

Now we are ready to commence our picture guide detailed in a step-by-step tour of a PV solar rooftop system installation. For the reader's convenience, the entire illustrated process is divided into four general "Photo Groups" and each "Group" is divided into several "Steps." Each "Step" is furtherdivided into several photos, all in chronological order of the installation process.

Site Evaluations

Step 1. Evaluating the electric service location

1. In the photo, we can see the installer has located the main service panel andin this instance we can see that he is in the process of installing a combination PV meter/main service panel.

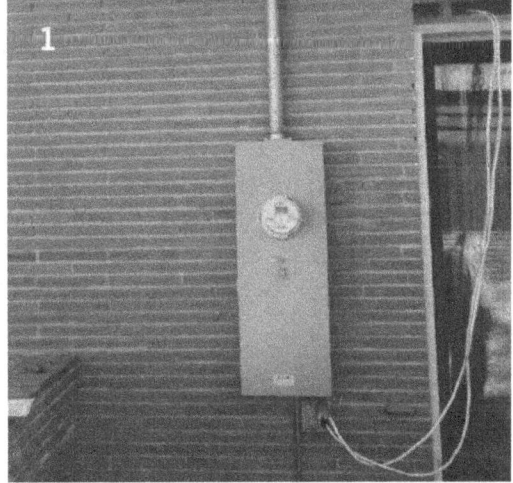

2. This photo shows the inside of the main service panel being installed and integrated with the utility meter. The installer is about to install as many internal circuit breakers as the home may require. Most will be 120V breakers but a few will be 220 to 240V breakers for certain appliances, suchas the electric water heater.

3. This photo shows the large label that contains all the specifications of the main panel. You will find this label affixed inside the main service panel door.

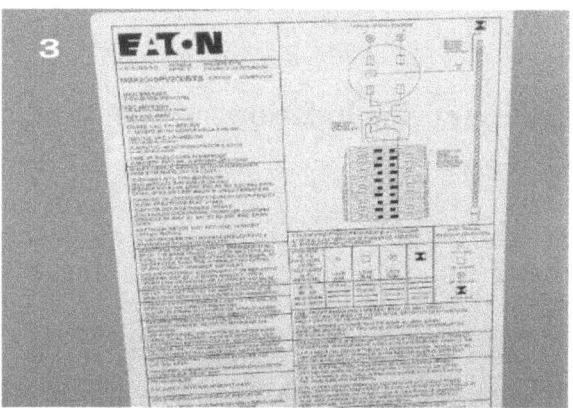

4. This photo is a close-up of the main breaker inside the main service panel. Notice that the switch is labeled with the overall amperage of the panel—in this case, 200 amps. In terms of electrical infrastructure, this is a fairly large home.

5. This photo shows the site evaluator removing the face plate of the main service panel. Since this is a new installation, he is in the process of inspecting the main panel to see that all connections are secured and tight.

Step 2: evaluate the structural capacity

6. In the photo, the installer is measuring the sub-roof structural beams. We can see by the tape measure that the width of these beams is a standard 9 inches, which is sometimes quoted as 10 inches.

7. In this photo, the installer is measuring the spacing between structural members. In this case it is a standard 16 inches between centers. With these two measurements we can easily find the load bearing capacity of this roof per square foot. To determine the load limits for specified open areas, we also need the length of the open beams between vertical supports. Standard load bearing tables will give us the answers. The lumber supply depot supplies tables with this load bearing information.

8. The information in this photo, namely the data printed on the side of the main beam, provides us with the basic truss identification and the manufacturer or the supplier/distributor can supply the load bearing tables.

Step 3: evaluate the shading

9. In the photo to the right, we can see the site evaluator is taking solmetric readings to determine the level of solar access. This is the key number for calculating the average seasonal solar production that can be expected from each PV module and from the total array.

You do not have to possess a solmetric measuring device. As explained in detail in subsection 5.6, you can obtain the insolation or irradiation index number for your precise location from the table provided, and also from the website identified in the same subsection. We explain how to use this irradiation index number, which gives you the average number of "sun hours" (summer average, winter average, and yearly average) for practically any location in North America.

Step 4: measure roof surfaces

10. In the photo to the right, we see the DIY installer is taking accurate measurements of the roof surface to calculate a simple plan for a potential layout for the PV array.

11. In the photo to the right, we see the DIY installer is taking and recording careful measurements from the edge of the roof to all the built-in roof obstructions. These measurements will enable the installer to prepare a drawing with an optimal and accurate layout for the panel array.

Prepare the Roof for Mounting the Solar Panels

Step 1: choose a general layout for the selected number of panels

The objective here is to avoid shading onto the PV panels from nearby trees or other rooftop structures (chimneys, satellite dishes, etc.).

1. This is a frame taken from a simple 3-D modeling software that enables the homeowner to evaluate the effects from possible shading on the roof surfaces at different times of the year caused by nearby trees, other buildings, chimneys, satellite dishes, and other roof structures. This can be very useful for choosing a general layout and configuration for how you will eventually mount your solar array.

2. The image here is a 3-D rendering of the solar array as it is proposed to be mounted of the south slope of the roof. It is prepared on a user-friendly software that allows the homeowner to get a bird's eye view of what a 27 module array would look like in different configurations on the rooftop before deciding on a final layout configuration. Spacing can easily be adjusted to allow for a chimney or other obstruction as may be required.

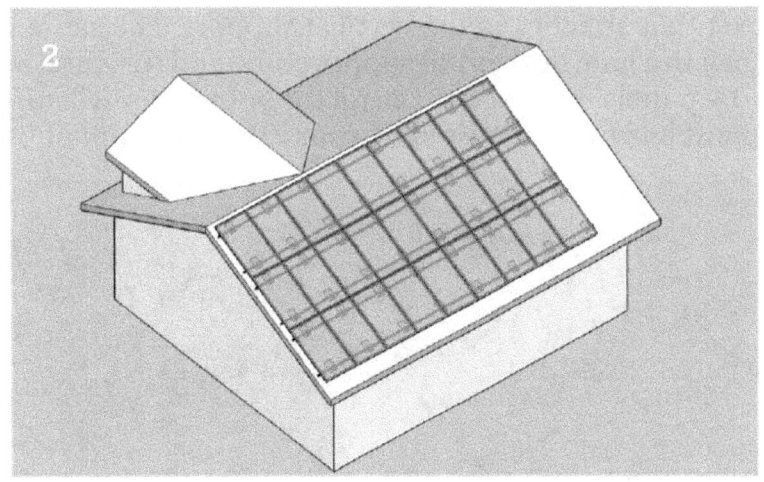

3. In the image on the next page, our DIY solar coach has prepared an interactive graphic that enables the homeowner to experiment with alternative layouts for the array. In this example, there are a total of 27 modules divided into two strings, or groups, one string of 13 modules at the lower section of the roof and the other string of 14 modules fits into the space immediately above the lower 13 PV panels. The model indicates the positioning of the rail supports under the array of modules. It also displays the location of the junction box (lower right) and conduit that will contain the DC wiring of the modules ("homeruns") and channel the wiring safely from above the roof and down through the roof to the AC disconnect switch located below on the side of the house under the eaves trough.

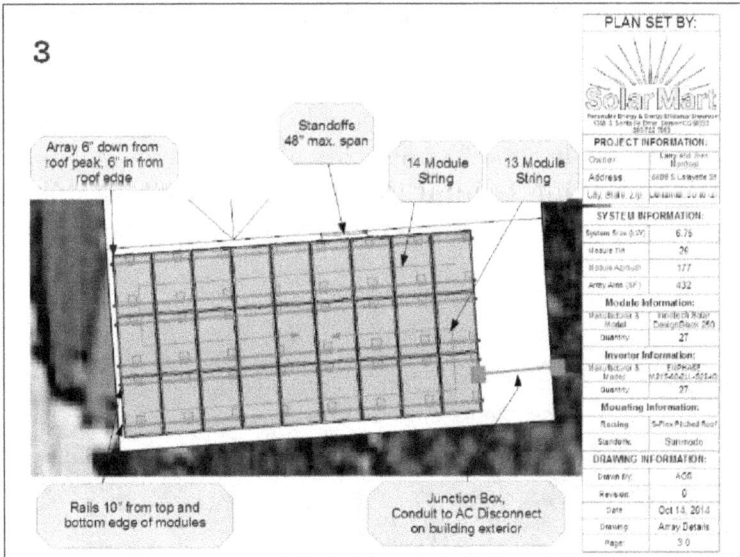

Step 2: selecting a suitable angle against the horizontal for the panel mounting support frames

4. In the photo, we see the DIY installer is in the process of choosing the best angle against the horizontal to enable the solar modules to optimize energy production. He does this by adjusting the height of the 3 vertical supports (that are each fixed at the bottom end of each support) to a fixed steel flashing bolted into one of the roof joists with lag bolts (see close-up photos below) and the flashing is fitted topside, above the shingles, with a

133

mounting bracket that attaches to the vertical support. The same vertical support at the top end is attached to a long horizontal support tube and this tube is attached to the upper section of the 4 PV panels. Your solar contractor or installer willsupply all of the specified items of the support hardware required by your PVpanel supplier, if you are using one.

The small array above of 4 modules may be additional to modules installed on the other side of the roof to increase total system production, or,on the other hand, the main array might be mounted on this side to avoid ashading problem on the other side.

5. In this photo, the installer has mounted the PV panels in an array on a planeparallel to (or flush with) the surface of the roof. Even if the flush mount does not give the optimum tilt angle, it is often used for ease of installation and aesthetic reasons. Lowering the tilt angle below the optimum angle will result in very minor energy losses. (See details in Chapter 5.) In the lower right-hand corner of the photo we can see the location where the homeowner decided to install the central inverter, disconnect switches, main distribution panel, metering device, *etc*. This is apparently at the back of the house besidethe living room windows. If you look closely you will also see the sky is partially overcast with sunshine at the top of the hill behind the house but with shade on the PV panels. So at this particular moment, we can assume that the PV solar rooftop system is not generating any power.

Step 3: choose the best available azimuth east–west angle for mounting the pv panels

6. In this photo, the DIY installer has just laid out the parallel markings (red lines) for the panel support mountings. The yellow 90-degree markings indicate the corners of a two-

module group. The yellow circles indicate the drill points where the lag screws will attach the steel flashing mounts to the roof joists just below the shingles. The exact location of the joists is revealed to the installer with the use of a quality electronic handheld stud-finder with the depth reading adjustment button set at maximum depth. This instrument is not expensive and it is accurate and reliable. The DIY installer will definitely want to buy one if he doesn't already own one.

7. In the photo, the installer is drilling a pilot hole for the flashing. The procedure to determine the exact locations for the drill holes and lag bolts is explained in detail in the subsection of the previous photo. For this job, if you are a DIY installer, you will need a good quality 1/2 HP portable electric drill with a spare rechargeable battery and a good set of drill bits if you don't already own these items.

STEP 4: PREPARING THE LAYOUT FOR THE PANEL SUPPORT FRAMEWORK

8. In the photo, the installer is prying up a shingle where the steel flashing will be inserted between shingles and secured with a lag bolt to a roof joist below the shingles to hold the mounting framework.

9. In the photo, we see the installer applying a clear silicon sealant in a double semicircle fashion to the back of the steel flashing. The sealant semicircles will be pointed downward when the flashing is in final position and this way water leakage through the roof at the point of the lag bolt is prevented.

Step 5: how mounting support hardware secures the panel framework to the roof

10. In the photo, the installer is applying the sealant to the pilot hole. This is a backup precaution in addition to the sealant applied to the underside of the steel flashing to prevent rainwater from getting into the hole. The hole in the center of the flashing will then be lined up with the pilot hole as shown in the following photo.

11. In the photo, we can see the DIY installer is fitting the steel flashing with a mounting bracket between the shingles as indicated by the red markings.

12. In the photo, we can see the installer has inserted the steel flashing into position between the shingles.

13. In the photo, we see the installer has positioned the lag screw ready to be screwed into the roof support framework joists immediately below the shingles. Drill holes are positioned with the use of a good quality "stud finder." The details of this procedure are discussed in the subsection of the photo above.

14. In the photo, our DIY installer is in the process of installing a lag screw into the roof support beam below the shingle. The lag screw will provide strength to the support framework rails that will be attached to one or several solar modules. (See additional photos below.)

15. In the photo, we can see one of the flashings that has just been installed.

16. In the photo, the installer has just prepared the steel support bracket for drilling through the flashing to the roof joist below. The joists below the shingles can be precisely located with a good quality large depth stud-finder.

17. In this photo, we can see the DIY installer is midway through installing one of the steel support brackets. Vertical supports will then be attached to a horizontal support railing. In this way, all the vertical supports will connect the horizontal railing to all the support brackets as illustrated in the photos below.

18. In the photo, we see the installer has just finished installing the steel support bracket on top of the steel flashing in line with the same red marking that lines up with the support brackets on either side of this one. This is how the vertical supports will align properly with the horizontal railings. Each PV solar module is attached with fittings called end-clamps or mid-clamps (see close-up photos below) to 2 horizontal railings, one about 20 centimeters from the top of the module and the other also at 20 centimeters from the bottom of the module. This means that each module or panel is attached firmly at 4 points around the outside of the panel to hold the panel tightly and securely, even in high wind conditions.

Step 6: how the panel mounting frameworkattaches to the solar panels

19. In this photo, our DIY installer has just attached the support hardware (90degree steel bracket) to the roof on top of the steel flashing with a steel spacer and he bolted the bracket onto the lower horizontal railing of the mounting frame support. The 2 horizontal railings (upper and lower) will provide the mounting support framework for the solar panels.

20. End-Clamp A close-up photo of an "end-clamp." Two end-clamps attach the outside edge of one of the PV modules to the support railing. Usually there are 2 or sometimes 3 rows of modules in an array, so this means there would normally be 4 or 6 end-clamps on the same railing on the left and the same number of end-clamps on the right where the array ends.

21. Mid-Clamp

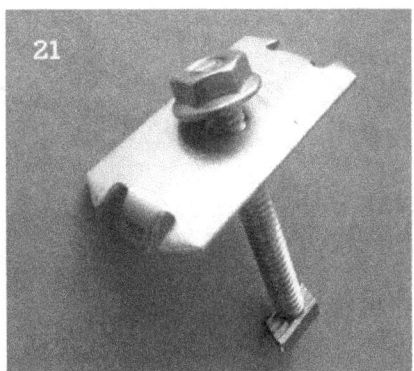

The other fitting shows a close-up photo of a "mid-clamp." There are usually 2 mid-clamps on the left and right of every module except for modules that are on the outside of the array, since the outside edge will be secured to the support railings logically with end-clamps, not mid-clamps.

We recommend that the DIYer purchase a few extra end-clamps and mid-clamps because a worker might accidentally kick one or two clamps lying on the roof into the garden and they won't be easily found. A little precaution here may save a lot of time not having to stop work to go to the suppliers shop to buy more clamps.

22. In the photo, the installer and his assistant are installing 2 end-clamps on the horizontal support railings, getting ready to secure an outside PV module.

23. In the photo on the next page, our DIY installer is using an open-end wrench to tighten up one of the 2 end-clamps (upper and lower), which will attach the PV panel outer frame to the horizontal support railing.

Also, in the photo, we can see that this PV installation is using monocrystalline silicon PV modules as opposed to polycrystalline silicon modules. The differences in appearance and performance and the various pros and cons of these two types of modules are discussed and illustrated in Chapter 1.

24. In the photo on the next page, the DIY installer has captured a good close-upshot of exactly how an end-clamp is attached to the support railing below and how the clamp secures the solar panel above to the support railing.

25. In the photo, we can see that the DIYer has installed one of 2 mid-clamps, getting ready to install the next panel parallel to the panel already installed.

26. In the photo, the installer is using a standard open-end wrench to tighten 2 mid-clamps holding in place 2 modules in perfect parallel position to one another.

Step 7: wiring the array through the roof to the electrical components

27. In the photo, the installer has just connected the wires (homeruns) from the PV panels together and he is feeding them down through the shingles of the roof through a waterproof steel and rubber fixture called a "weather-head."

The homeruns are then connected to a DC disconnect switch and then to the DC/AC inverter installed on the outside of the house in the electrical equipment area. More details are provided in photo groups 3 and 4.

Mounting the Solar Panels

Step 1: how to get the pv panels safely up on the roof

1. In the photo, the DIY homeowner and his assistant are manually uploading a solar panel from the man on the ladder (there is another helper at the bottom of the ladder). In this

manner, each PV module can be safely and manually transferred from the ground level or the driveway up to the roof.

2. In the photo, the DIY installer's assistants are helping the delivery truck and crane to unload the solar PV modules and rails at the job site. These workers should be wearing construction hard hats and for safety should avoid standing directly under the load.

3. In the photo, we see the DIY installer and his crew guiding the crane driver and unloading the PV module packages onto the roof from the truck below. Notice that the horizontal support railings have already been installed and provide a bracing to prevent the module packages from sliding down or falling to the ground. A precautionary note: Be careful not to unload your module bundles in an unsafe position or on a dangerous slope of the roof where they might slide down and get damaged.

This is a fairly large installation but for smaller systems, the installer and his crew will often manually raise the PV modules up to the roof one by one using a ladder.

Step 2: safety precautions

4. In the photo, we see one of the installer crew using the recommended safety harness with a safety rope firmly attached at the peak of the roof. When handling fragile, heavy loads, you must always think "safety first." Not only for the safety of the PV glass and aluminum modules, but also for the safety of yourself (if you are a DIY installer) and for the whole installation crew.

5. In the photo, we can see that the DIYer is using an insulated fiberglass ladder that provides another measure of safety. Standard aluminum ladders are not recommended since they can act as electrical conductors. Once again, think "safety first!"

6. The photo demonstrates the kind of steel-toed standard safety boot recommended to be worn at the job site by all members of the installation crew.

Step 3: installing pv panels to support theframe

7. In the photo, the DIY installer and his assistant are in the process of installing the first PV module after they have installed the horizontal supportrails. You will observe that the installer is not wearing his safety harness, acommon lazy habit we recommend you avoid. "Safety first" should alwaysbe the number one rule for crews working on rooftops.

8. In the photo, we can see the DIY installation crew is in the process of installing the first PV module of the array on the bottom right hand corner ofthe roof. An observant reader may notice that the horizontal support railings are installed parallel with the bottom edge of the roof and that they are alsosecured equidistant from both the top and bottom edge of the module. Whenfirmly installed, this mounting arrangement gives optimum strength for themodules to resist high wind conditions without damage.

9. In the photo, the DIY installer has selected the approximate position for installing the end module at the opposite end of the bottom row of the array.

10. In the photo, we can see that the install crew is getting ready to attach a "tension string marker" for the purpose of aligning the bottom row of modules. This process is continued in the next photo.

11. Here, in the photo, we can see the installer has tightened the alignment stringmarker the entire length of the bottom row, which will facilitate aligning all the panels in the bottom row to be parallel with the edge of the roof and parallel with each other.

12. In the photo, we can see that the install crew has progressed and is now lining up and installing module (approximately) number 9, working from right to left. By taking extra care in the proper alignment of the bottom row of PV modules, the DIY installer is ensuring that the second and third rows will also be perfectly aligned and this will give an attractive finish to the final array that would make any homeowner proud of his PV solar system.

Step 4: connecting the pv panels (modules)

13. The photo is a close-up shot of a female connector used to interconnect the modules together. These are used with quick-connect male connectors (as shown in the following

two photos).

14. In the photo, we can see that the install crew has just mounted an upper railing to which they are attaching the PV DC conductor wires or homeruns between the modules. These wires are secured to the railing with plastic tensor strips to resist high winds and will not be visible after the modules are installed, but the male–female connections will remain accessible for service if needed.

15. This close-up photo shows a male and female connection that have been attached to one another and this line is labeled as part of "string 1" that includes the 13 modules of the bottom row shown in the photos above.

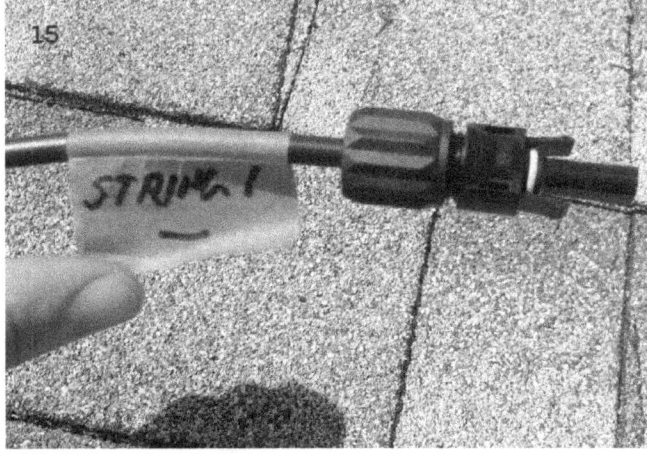

16. The photo shows how the cables or leads arrive from the manufacturer, carefully taped to the back of the panel so they're easy to free up when the installer is ready to interconnect the modules into a "string." It appears that our DIY installer is about to install and connect this PV panel.

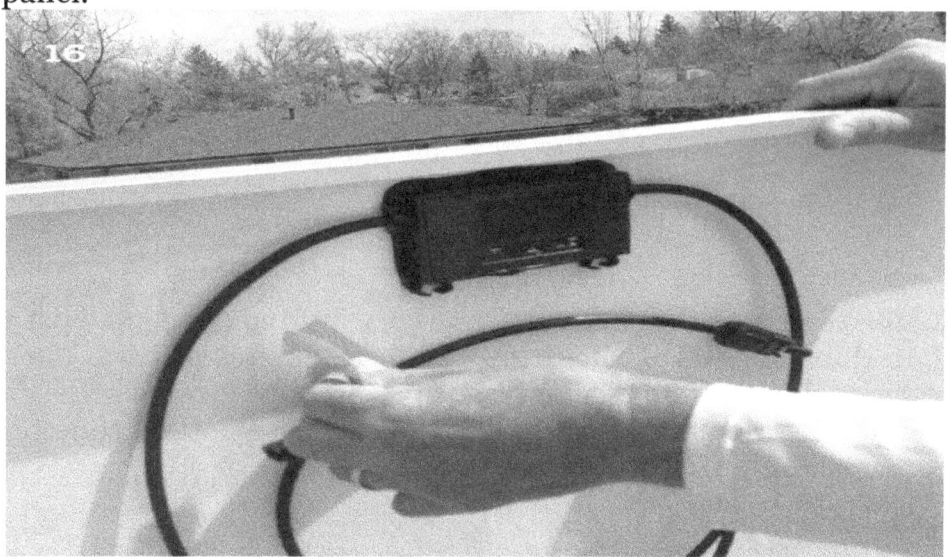

Step 5: how solar panels are wired together

17. The photo shows the DIY homeowner or his helper in the midst of wiring the first 3 PV modules of the bottom row of the array.

18. As shown here, it can be a bit of a tight fit to make the final connection between the leads of adjacent PV modules but the mounting support frames are engineered to allow a normal size person or worker to reach under the modules and complete the task of wiring the system.

Installing the Electrical Components and Controls

Step 1: installing the conduit from the array

1. This is a close-up photo showing how the installer has just attached a junction box at the end of a string of PV modules (the array). The junction box simply unites several PV production wires or homeruns from the modules and combines them to run through the Electrical Metal Tubing (EMT) commonly referred to as the conduit, which then passes through a waterproof fitting from above the roof to below the roof where the wires will be connected to the electrical controls in the equipment area.

2. The photo is taken from the other side of the junction box and it shows how the DIY installer can use the conduit and a 90-degree conduit fitting to penetrate the roof. Notice the use of silicon to weatherproof the holes. The alternative is to feed the homeruns directly through a weather-head.

3. The photo shows how the DIY-install crew makes a 90-degree bend in the conduit tubing without the need to connect or disconnect interior wires. This part of the conduit is held in place by a steel flashing, clamp, and lag bolt.

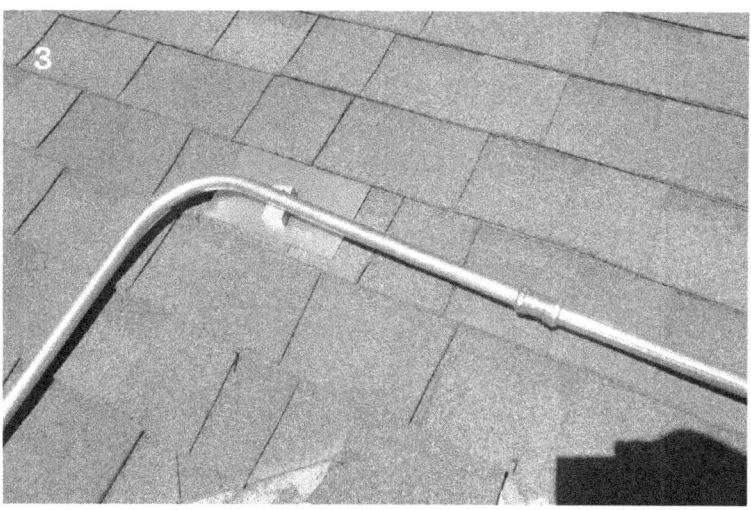

4. In the photo, we can see the DIY-install crew is attaching the conduit to a flashing with a lag bolt, spacer, and clamp.

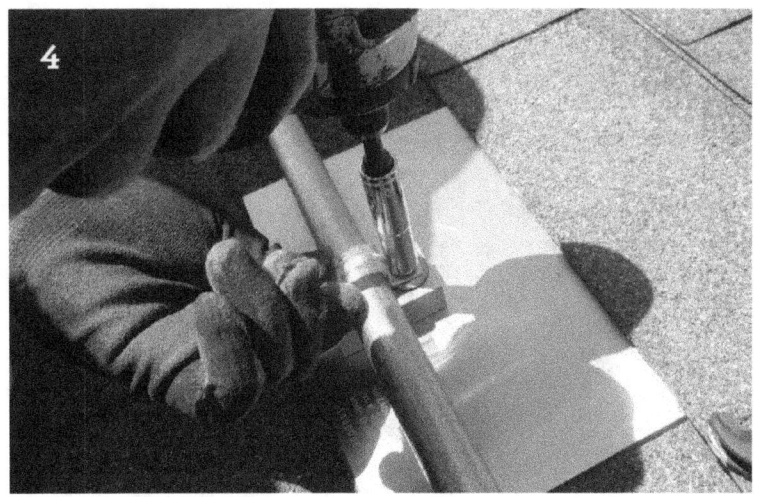

Step 2: install dc disconnect switch before the central inverter

5. The photo shows a DC disconnect switch installed after the PV panels and just before the central inverter. When activated, this switch will isolate the solar modules when the inverter needs inspection or servicing, or for adding additional modules or other modifications to the system.

6. The photo shows the inside of the DC disconnect switch box. This switch is ready to be installed between the array modules and the central inverter.

7. In the photo, the DIY installer has just connected and wired a DC disconnect switch under the "Sunny Boy" central inverter.

8. In the photo, we see that the installer has removed the cover plate from the DC disconnect switch and we can see all of the wiring including the 4 (black) leads from the conduit coming from the module string. At the top of the switch box, we can see the red wires leading into the central inverter.

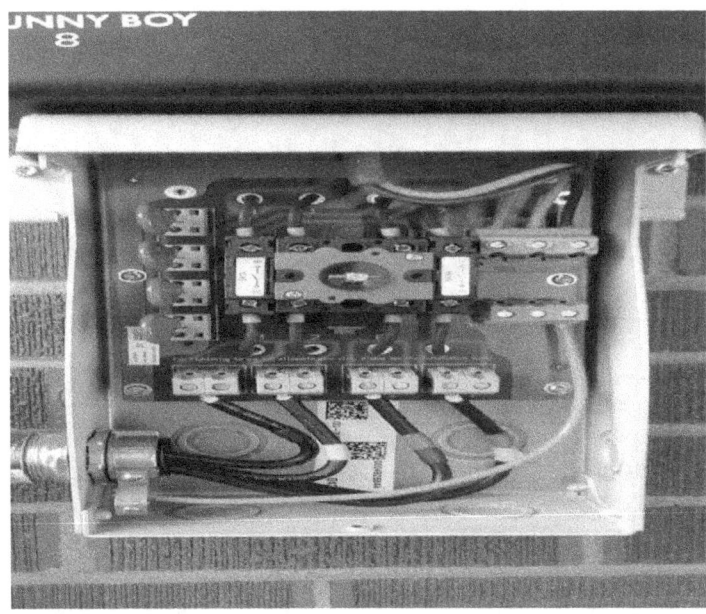

9. The photo is a good close-up of the 4 black wire homeruns entering through the conduit into the box and attached to the 4 green terminals of the DC disconnect switch. A standard good quality straight-blade small screwdriver is used to supply sufficient torque to fasten the homeruns.

Step 3: install the central inverter

10. In the photo, the installer has just attached an inverter attachment fixture to the wall with an easy disconnect bracket in order to detach the central inverter for service.

11. In the photo, the inverter is hung on the wall attachment bracket but has not yet been wired.

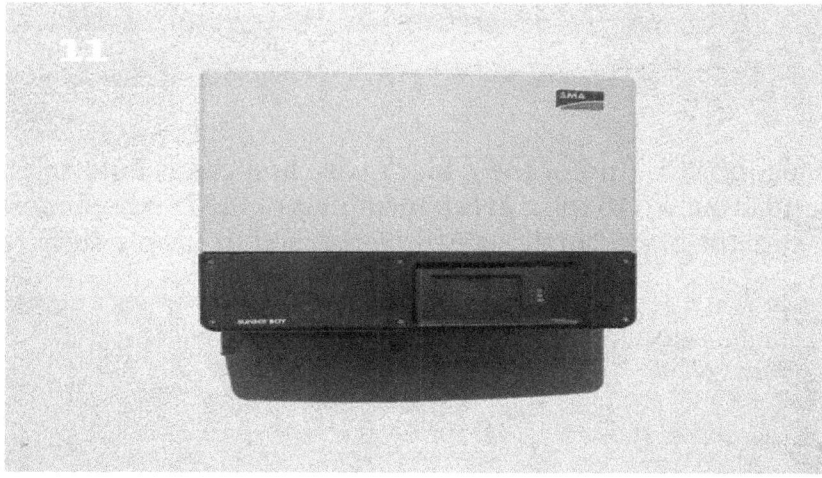

Step 4: install micro-inverters (alternative to a central inverter)

Micro-inverters are an excellent alternative to central inverters. The basic differences as well as the advantages and disadvantages of each type of inverter are explained in detail in Chapter 4.

12. This is a close-up photo of a standard Enphase 215 watt micro-inverter being attached to the upper railing before installing the corresponding PV module. This system is being installed with micro-inverters instead of a central inverter. Micro-inverters have been experiencing considerable success since their introduction into the market a few years ago. They have significant technological advantage (among other advantages) over central inverters and these advantages are explained in considerable detail in Chapter 4.

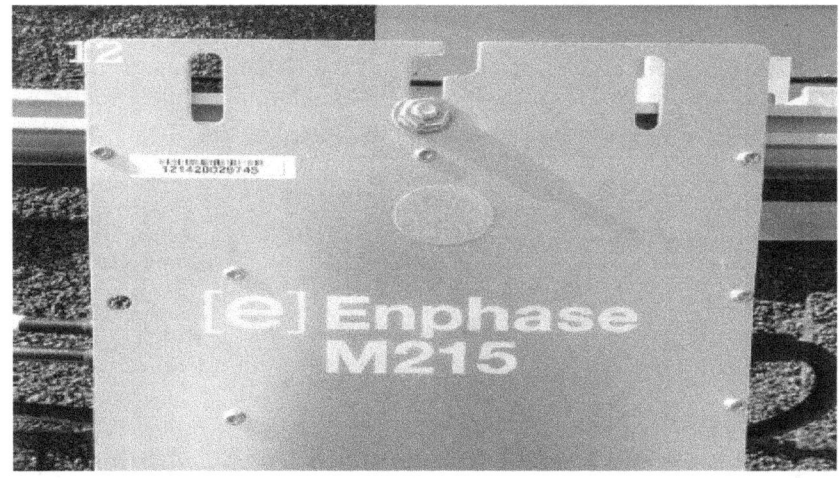

13. The photo shows a small micro-inverter with its 2 leads ready for connection. When micro-inverters are incorporated into each PV panels, youcan find them installed on the backside of each module along the upper edgeright-hand corner and they are not visible after installation from the side orfrom above.

Step 5: install ac combiner panel and ac

DISCONNECT SWITCH

14. In the photo, several electrical devices are being connected in series, including: (L to R) the main service panel, the AC disconnect switch (this is separate from the DC disconnect switch), the photovoltaic system productionmeter, and the AC combiner panel.

15. The photo shows a close-up shot of the inside of the AC combiner panel with the wiring coming from the PV production meter (through the conduit at the bottom left side of the photo) and, after being combined in the box, the wiring carrying the current exits through the other conduit on the lower right side.

16. This photo shows the AC combiner panel all wired up with its face plate attached.

17. The photo shows an AC disconnect switch in the process of being installed.

Step 6: install electrical gutter box

18. The photo shows how the DIY installer crew has just installed an electricalgutter box that combines the output AC lines from the 2 inverters above.

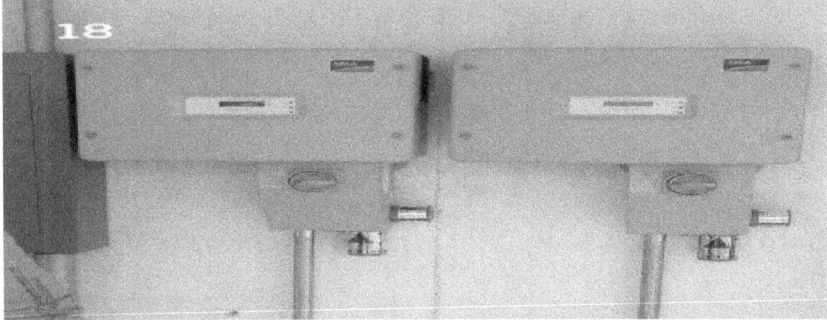

Step 7: install utility net meter

19. The photo shows the next step of the installation, which is the utility net meter. This meter charges the house account with the electricity consumed asreceived from the utility but it also credits the same account with any excess electricity generated from the PV system that is not used by the home; rather,it is fed back into the grid. The sum of these credits during the month can reduce your net electric utility bill significantly.

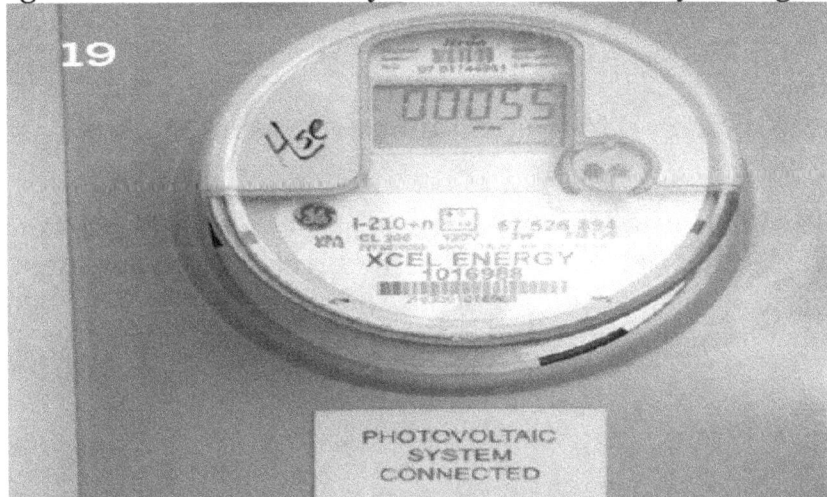

Step 8: install the utility net meter and the pvproduction meter in a separate box

20. The above photo shows a large panel combining a utility meter with the mainservice panel. They can be together in the same panel but are often installed separately. The PV production meter (a separate meter) is then connected to the main service panel.

21. The above photo shows a close-up shot of a house waiting to be fitted with a photovoltaic production meter.

Step 9: install home main service panel and pvbreaker switches

12. The above photo shows that the DIY installer is beginning to install the mainservice panel and we can see he has started to label some breakers for easy identification of circuits when anyone is inspecting or servicing the main panel. The hand-written labeling identifies which breakers control which circuits or appliances.

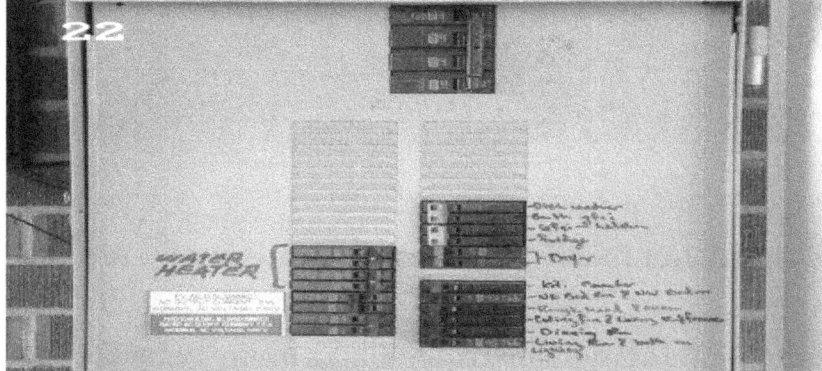

13. The above photo shows how the installer has used two different breakers to connect to two separate PV circuits. For example, and for convenience, one breaker might control the lighting of a certain area of the house and the other breaker might protect specific appliances that need electric backup in the event of a grid blackout.

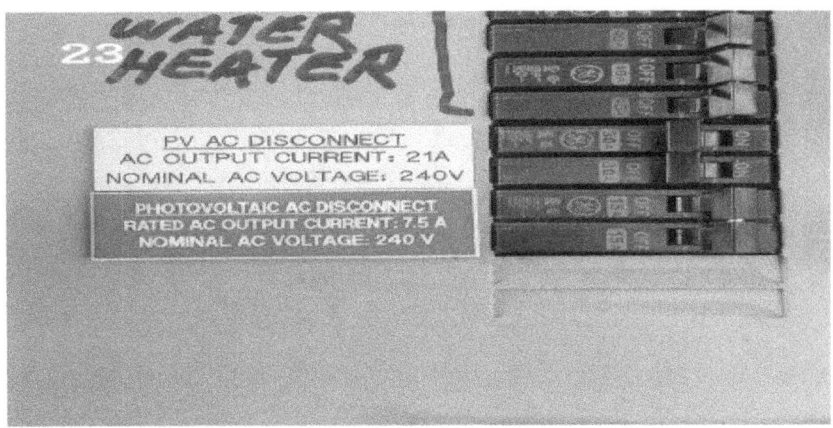

Step 10: install surge protector

14. This is a close-up photo of a panel that has just had a surge protector installed.

Step 11: install monitoring device (optional but recommended)

15. In the above photo, we see the DIY-install crew has installed a final optional device known as a "monitoring gateway." This device will permit the homeowner and any service personnel to monitor the production performance of all the individual PV modules in the system and can help locate any defects that might need attention.

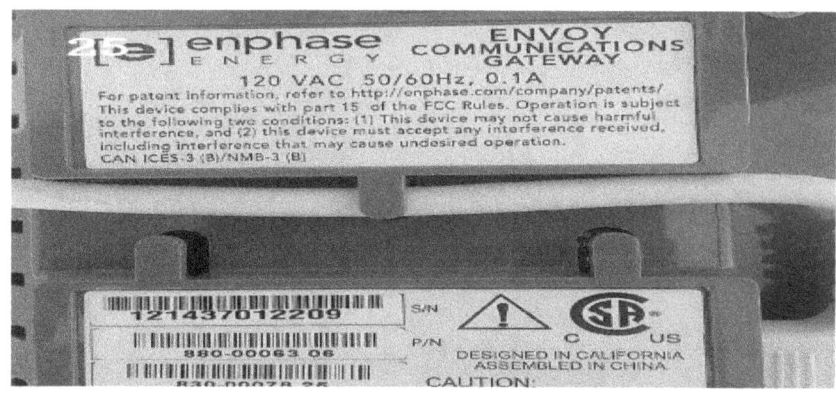

Step 12: how it all fits together

16. The above photo shows a relatively simple system where we can see the equipment area with the PV system equipment items installed. From left to right, the DIY-install crew has installed in series the DC disconnect switch that is connected directly into the bottom left of a central inverter; the inverter is then connected to the AC disconnect switch and then to the PV production meter and then it is connected to a combined utility net meter and the main service panel. Note that the PV production meter housing is installed with conduits and wires but the PV meter has not yet been installed.

17. The photo shows a second system that is slightly larger than the system shown in the previous photo because it has two central inverters instead of one.

18. The photo shows the setup of the equipment area of a third, relatively simple PV solar system. Most of the same equipment items in the previous two systems are repeated here with only a few exceptions, but in this installation the current will flow from right to left.

Conclusions and Suggestions

I trust that this book has helped demystify some of the vague unknowns and confusing propaganda about PV solar energy systems, and that it has shown you the basics of planning and properly installing an efficient PV solar rooftop system.

You've seen the layout of a system from the rooftop panels to the grid, and now you understand what components you'll need to buy and how to make intelligent decisions about choosing the optimal type and number of PV modules, selecting the best type of inverter for your system, and deciding on the type of mounting support system and how to determine the best layout of the PV panels for your particular roof.

Whether your goal is to save a lot of money on your electric bills, to generate your own clean energy, to increase the appraised value of your home, or all of the above, then investing in a PV solar rooftop electric system is a wise decision. Even a relatively small solar electric system can produce reliable, economical, pollution-free energy for your home.

A PV solar system that can provide 100 percent of your total consumption requirements might seem appealing, but it may not be financially feasible for your budget. Or it may not be practical due to space limitations on your rooftop. As a general rule, about 100 square feet (equivalent to five solar panels) will generate roughly one kilowatt per hour (kW/h) of electricity, but only when the sun is high and perpendicular to the panels. Most residential solar electric systems require at least 100 to 200 square feet (for small "starter" systems), and up to 1,000 square feet for 40 to 50 modules. Of course, commercial and industrial PV rooftop systems require much more surface area than this.

Regardless of the PV system size, residential or commercial, the basic contents of this book will still apply.

The total installed cost of a PV solar system is a considerable investment, but the overall cost is getting cheaper by the year, and in many countries the cost of PV power for the homeowner has already reached the goal of "grid parity." Grid parity is when the cost of producing electricity by PV solar is equal to the cost of conventional electricity produced by large generators that burn carbon-based fuels. PV solar-generated power becomes more attractive and more competitive as the electric utility billing rates go up. Moreover, since the investment payback period is usually only five or six years with government financial incentives, and since the electricity cost savings continue for another 20 years after payback (electricity essentially becomes free after the payback period), PV solar makes eminently good business sense, both for the homeowner and the banks that finance PV rooftop systems.

Some people have a flawed concept of solar energy. While it's true that sunlight is free, the electricity generated by PV systems is not. There are many factors involved in determining whether installing a PV system would be an economical alternative for any given home. There's the question of geographicallocation, installation cost, and true personal long-term benefits relating to yourunique lifestyle and needs.

However, the demand for PV solar continues to grow as equipment prices fall, to the point where PV solar systems produce power at competitive levels (grid parity) with conventional power from the utility company. The demand for PV solar is also stimulated as the world becomes increasingly aware of the environmental concerns associated with conventional power sources, which depend on burning carbon-based fuels. Considering all these growing trends worldwide, there's little doubt that photovoltaics have a bright future and will continue to grow.

Glossary of Terms

Ampere: Unit used in measuring intensity of flow of electricity. Its symbol is "I."

Alternating Current (AC): Electric current that reverses its direction of flow at regular intervals. For example, a current in a 60-cycle system would alternate 60 times every second. This type of current is commonly found in homes, apartment buildings, and businesses.

Bare Conductor: Wire or cable with no insulation or covering.

Current: Flow of electricity through a circuit; either AC or DC.

Circuit: Flow of electricity through two or more wires from the supply source to one or more outlets and back to the source.

Circuit Breaker: Safety device used to break the flow of electricity by opening the circuit automatically in the event of overloading; also used to open or close it manually.

Conductor: Any substance capable of conveying an electric current. In the home, copper wire is usually used.

Covered Conductor: Wire or cable covered with one or more layers of insulation.

Conductor Gauge: Numerical system used to label electric conductor sizes, stated as American Wire Gauge (AWG), with wire measured in square millimeters diameter.

Cable: Conductors insulated from one another.

Direct Current (DC): Electric current flowing in one direction. This type of current is commonly found in manufacturing industries. It's also the type of current produced by solar panels, which is then converted by an inverter into usable alternating current.

Electricity: Energy used to run household appliances and industrial machinery; can produce light, sound, heat, and has many other uses.

Frequency: The number of periods per unit time stated in cycles per seconds, or Hertz. For alternating current power lines, the most widely used frequencies are 60 and 50 Hertz.

Fuse: Safety device that cuts off the flow of electricity when the current flowing through exceeds the fuse's rated capacity.

Ground: To connect with the earth, as in grounding an electric wire directly to the earth or indirectly through a water pipe or some other conductor. Usually, a green-colored wire is used for grounding the whole electrical system to the earth. A white wire is usually used to ground individual electrical components.

Impedance: A measure of the complex resistive and reactive attributes of a component in an alternating current circuit.

Insulator: Material that will not permit the passage of electricity.

Inverter: An electric devise that converts direct current (DC) into alternating current (AC).

Neutral Wire: The third wire in a three wire distribution circuit. It's usually white or light gray and is connected indirectly to the ground.

Resistance: Restricts the flow of current; the unit of resistance is called an "Ohm." The more the resistance in Ohms, the less the current will flow.

Service: Conductor plus the equipment needed to deliver energy from the electricity supply system to the wiring system of the home or premises.

Service Drop: Overhead service connectors from the last pole connecting to the service conductors at the building or other electricity consumption destination.

Service Panel: Main panel or cabinet through which electricity is brought into a house or building and distributed. It contains the main disconnect switch and fuses or circuit breakers.

Short Circuit: Break in the flow of electricity due to excessive current, resulting from a fault or negligible impedance between live conductors having a difference in potential under normal operating conditions.

Voltage Drop: Voltage loss when wires carry current. The longer the cable or chord, the greater the voltage drop.

Volt: Unit for measuring electrical pressure or force, known as electromotive force. The symbol for volt is "E" or "V."

Wires: Conductors carrying the electric current or power to the load; usually black or red.

Watts: Unit of electric power, calculated by multiplying volts by amperes.

www.ingramcontent.com/pod-product-compliance
Lightning Source LLC
Chambersburg PA
CBHW081111080526
44587CB00021B/3545